Climate Activism
in the 21st Century

T0197885

ALSO BY DIANNE RAHM
AND FROM MCFARLAND

*Climate Change Policy in the United States: The Science,
the Politics and the Prospects for Change* (2010)

*Sustainable Energy and the States: Essays on Politics,
Markets and Leadership* (edited by Dianne Rahm, 2006)

*Toxic Waste and Environmental Policy in the 21st Century
United States* (edited by Dianne Rahm, 2002)

Climate Activism in the 21st Century

DIANNE RAHM

McFarland & Company, Inc., Publishers
Jefferson, North Carolina

ISBN (print) 978-1-4766-8534-2
ISBN (ebook) 978-1-4766-5068-5

LIBRARY OF CONGRESS AND BRITISH LIBRARY
CATALOGUING DATA ARE AVAILABLE

Library of Congress Control Number 2023032584

Front cover image: © Jacob Lund/Shutterstock

Printed in the United States of America

*McFarland & Company, Inc., Publishers
Box 611, Jefferson, North Carolina 28640
www.mcfarlandpub.com*

*To Anastasia and Jay
with deep thanks for their loving support.*

Table of Contents

Acronyms
and Initialisms Used

ACE—Action for Climate Empowerment
AIM Act—American Innovation and Manufacturing Act
ALEC—American Legislative Exchange Council
ALF—Animal Liberation Front
ANWR—Arctic National Wildlife Refuge
AR4—Fourth Assessment Report
AR6—Sixth Assessment Report
ARRA—American Recovery and Reinvestment Act
BAU—Business as Usual
BINGOs—Business and Industry NGOs
BLM—Bureau of Land Management
CAFE—Corporate Average Fuel Economy
CAP—Climate Action Plan
CARB—California Air Resources Board
CARES Act—Coronavirus Aid, Relief, and Economic Security Act
CBD—Convention on Biodiversity
CCP—Cities for Climate Protection
CCS—Carbon Capture and Storage
CDC—Centers for Disease Control and Prevention
CDM—Clean Development Mechanism
CDR—Carbon Dioxide Removal
CEI—Competitive Enterprise Institute
CEQ—Council on Environmental Quality
CFCs—Chlorofluorocarbons
CH^4—Methane
CHIPS—Creating Helpful Incentives to Produce Semiconductors and
    Science Act
CO^2—Carbon Dioxide
CO^2e—Carbon Dioxide Equivalent

COP—Conference of Parties
Covid-19—Coronavirus disease 2019
COY—Conference of Youth
CPP—Clean Power Plan
CSD—Commission on Sustainable Development
DACA—Deferred Action for Childhood Arrivals
DEQ—Department of Environmental Quality
DOD—Department of Defense
DOE—Department of Energy
DOI—Department of the Interior
DOT—Department of Transportation
ECOSOC—United Nations Economic and Social Council
EECBG—Energy Efficiency and Conservation Block Grant
ELF—Earth Liberation Front
EPA—Environmental Protection Agency
EU—ETS European Union Emissions Trading Scheme
EU—European Union
EU-15—European Union 15
EV—Electric Vehicle
FAO—Food and Agricultural Organization
FAR—First Assessment Report
FFI—Fauna and Flora International
G7—Group of Seven
GCC—Global Climate Coalition
GCST—Global Climate Science Team
GDP—Gross Domestic Product
GEF—Global Environmental Facility
GHG—Greenhouse Gas
GMO—Genetically Modified Organism
GND—Green New Deal
HFCs—Hydrofluorocarbons
HLPF—High-Level Political Forum
ICLEI—International Council of Local Environmental Initiatives
ICT—Information and Computer Technology
IIJA—Infrastructure Investment and Jobs Act
IMF—International Monetary Fund
INDC—Intended Nationally Determined Contributions
IPCC—Intergovernmental Panel on Climate Change
IRA—Inflation Reduction Act
IUCN—International Union for the Conservation of Nature
JI—Joint Implementation
JMA—Joint Mitigation and Adaptation

LBGTQ—Lesbian, Bisexual, Gay, Transgender, Queer
LNG—Liquid Natural Gas
LULUCF—Land Use, Land Use Change and Forestry
MAGA—Make America Great Again
MEAs—Multilateral Environmental Agreements
MGGRA—Midwestern Greenhouse Gas Reduction Accord
MOU—Memorandum of understanding
N^2O—Nitrous Oxide
NAFTA—North America Free Trade Agreement
NAS—National Academy of Sciences
NASA—National Aeronautics and Space Administration
NCA—National Climate Assessment
NCEAS—National Center for Ecological Analyses and Synthesis
NDC—Nationally Determined Contributions
NEP—National Energy Policy
NEPA—National Environmental Policy Act
NF^3—Nitrogen Trifluoride
NFI—Nature Friends International
NGOs—Non-Governmental Organizations
NHTSA—National Highway Traffic Safety Administration
NOAA—National Oceanic and Atmospheric Administration
NRDC—Natural Resources Defense Fund
NSA—National Security Agency
OECD—Organization for Economic Cooperation and Development
OMB—Office of Management and Budget
OPEC—Organization of Petroleum Exporting Nations
OSTP—Office of Science and Technology Policy
PFAS—Poly- and Perfluoroalkyl
PFCs—Perfluorocarbons
PPE—Personal Protective Equipment
PTP—Powering the Plains
R&D—Research and Development
REC—Renewable Energy Certificates
REDD—Reducing Emission from Deforestation and Forest
   Degradation
RGGI—Regional Greenhouse Gas Initiative
SALT—Strategic Arms Limitation Talks/Treaties
SAR—Second Assessment Report
SCC—Social Cost of Carbon
SES—Socio-Economic Status
SF^6—Sulfur Hexafluoride
SGIG—Small Grid Investment Grant

SR—Scientist Rebellion
SRM—Solar Radiation Management
START—Strategic Arms Reduction Treaty
SWCCI—Southwest Climate Change Initiative
TAR—Third Assessment Report
TARP—Troubled Assets Relief Program
TCI—Transportation and Climate Initiative
TPP—Trans-Pacific Partnership
UN—United Nations
UNDP—United Nations Development Programme
UNEP—United Nations Environmental Programme
UNFCCC—United Nations Framework Convention on Climate
    Change
USD—US Dollars
USDA—United States Department of Agriculture
USSR—Union of Soviet Socialist Republics
UUC—United Church of Christ
WCI—Western Climate Initiative
WCS—Wildlife Conservation Society
WGA—Western Governors' Association
WHO—World Health Organization
WMO—World Meteorological Organization
WREGIS—Western Renewable Energy Generation Information System
WTO—World Trade Organization
WWA—World Weather Attribution
WWF—World Wildlife Fund
XR—Extinction Rebellion
YONGO—Official Youth-Constituency to the UNFCCC

Preface

Climate activism changed considerably in the first decades of the 21st century. The shifts were due to several factors. The framing of the climate issue moved from being one described mostly by experts and a narrow attentive public as one with mostly future impacts, to one understood by popular majorities as current and urgent. The reframing of the issue from climate change to climate crisis set the stage for a new set of activists to drive the issue forward. Many of these activists were youths who saw the climate crisis as an issue that would dominate their lives and deprive them of a good future. As the youth movement garnered popular attention, many adult supporters joined the popular demand for climate action. The rise of a world protest movement pushed the climate crisis onto the mainstream agenda in a way it had not been in the 20th century. Public and private sector leaders were forced to address the issue in new ways.

Over the first decades of the 21st century, new international agreements detailing how the world would respond to the climate crisis were negotiated. These new agreements departed from 20th-century agreements in important ways. The most important change was a reconsideration of national responsibilities for taking climate action. In the 20th century, responsibility for action was limited to developed nations, which were rightly seen as the polluters who caused the problem. By the 21st century the situation had altered. Previously exempted nations had now become part of the group of nations that were emitting greenhouse gases in large amounts. To solve the problem, all nations had to be brought into the agreement to limit emissions. Successfully negotiating this new direction proved challenging and raised some ancillary issues.

The meaning of climate justice changed in the 21st century. Prior to the turn of the century, scientists had been unable to attribute specific weather events to global warming but in the first decades of the 21st century that changed. The climate debate took place when destructive weather events could be directly attributed to global warming. While global leaders had long recognized that poor nations would need assistance from the affluent nations to adapt to climate change, as specific instances of damage and loss became clear, the demands grew beyond adaptation assistance to reparations. Climate justice was also further described as intergenerational injustice. Youth argued that their futures were being sacrificed because of the failure of the older generation to take meaningful action. The youth movement also widened its definition of justice to include many of the arguments long emphasized by developing countries where injustice was tied to colonialism and racism. These themes became part of the new climate justice movement.

While a persistent few clung to climate science denialism in the 21st century, that position largely shifted to delay and obstruction. The fossil fuel sector, allied corporations unwilling to give up their profitable business practices, and nations with a fossil fuel resource base they want to continue to profit from no longer deny the reality of the warming. Rather they put their efforts into finding ways to continue the status quo, albeit with a nod to the goal of moving to a lower-carbon future. While more radical activists push the demand for a full oil, gas, and coal phaseout, the powerful forces that profit from the current state of affairs seek to put on the brakes. Pledges of net-zero by mid-century have come from many of the polluters even as achievement of those goals remained beset with questions of accountability, transparency, and adoption of yet unproven new technologies to capture and store carbon dioxide emissions.

Awareness of the costs associated with the climate crisis grew substantially in the first decades of the 21st century as climate-related disasters multiplied. The pace of these disasters will likely grow in the next several decades until the world successfully transitions to a new low-carbon economy. But the first decades of the 21st century did mark the beginning of the transition to a clean energy economy. The rapid expansion of solar and wind energy along with the launch of the electric vehicle revolution signals a vast change. Managing that transition will be one of the major challenges of the next several decades. Opposition forces will continue to try to delay or stop the movement away from fossil fuels. Activists will have to carefully navigate progress toward their goals of a sustainable and livable planet.

This book provides an introductory treatment of each of these major themes that defined climate crisis activism in the first decades of the 21st century. The book begins with a discussion of the reframing of the issue to one that was defined by crisis and urgency. Chapter 2 discusses the many actors both within the U.S. and internationally that play a part in 21st-century climate debates. Chapter 3 reviews climate policy from the 20th century to set the stage for subsequent chapters that detail actions taken in the Bush, Obama, Trump, and Biden administrations. The recounting of these ends with the mid-term elections during the Biden administration. Chapter 8 discusses the centrality of the climate justice movement to 21st-century activism. The final chapter looks at emerging issues that will likely affect the climate crisis debate going forward.

CHAPTER 1

Framing the Climate Crisis
Policy Debate

Introduction

The concept of frames or framing has a long-established role in public policy debates. How a policy issue is conceptualized and presented can have a profound effect on what, if any, action is taken on a policy issue and how the public perceives the issue. Framing is highly significant in setting the policy agenda and defining problems, but it is also important in later stages of the policy process as stakeholders forge solutions to the defined problem.

The debate over our warming climate has passed through a series of frames, counter-frames, and reframing efforts. As the issue has been perceived to be more urgent the framing moved from an initial focus on global warming with mitigation as the suggested solution to climate change with adaptation and mitigation as the proposed policy solutions. As scientists grew more concerned that time was running out for the world to avoid the likely catastrophic impacts of warming exceeding 1.5 to 2 degrees Celsius by the end of the century, a sense of urgency reframed the debate. The new frame was the climate crisis or climate emergency and more extreme solutions such as geoengineering came under serious consideration.

Framing policy issues in urgent ways may be helpful in spurring action; however, the introduction of urgency may also result in permanent inaction if the public grows numb to the failure to solve problems even when they are framed as crises. The takeaway in that instance may be just the reverse of what framers seek if the response to emergency framing is lackluster. Time-pressured framing may also produce untoward results especially if crisis framing results in the narrowing of options considered useful to solve the problem.

After first exploring some of the details of framing in policy debates, this chapter investigates the transition in the climate debate from the original framing of global warming to climate change. The chapter next discusses a second reframing from climate change to climate crisis or climate emergency or another variant expressing urgency like climate chaos, climate disaster, or climate apocalypse. The counter-frame climate denialism and the loss-frame of economic cost are explored. Drawing on examples of two other policy issues framed as urgent, the chapter seeks to suggest the likely outcome of crisis framing for the climate debate.

Framing Policy Issues

While used by many academic disciplines, the concept of frames or framing has a long history in public policy studies going back to the 1950s. This early research looked at political communication in newspapers, radio, speeches, and news magazines. By studying the general "themes" (frames) raised, scholars thought it was possible to learn something about the general state of politics. Generally, these studies used a qualitative form of content analysis to distill the main themes or frames used in the policy debate. Policy frames are a form of a story or narrative that is told by political actors to define problems and set the political agenda (Nie 2003). Framing involves positioning messages in ways that emphasize certain aspects of the issue that political actors want highlighted and deflecting attention from aspects of the issue that political actors want ignored (Takach 2019).

Framing is a process in which political actors construct meaning by asking the question "Just what is it that's going on here?" By answering this question, they can make sense of what is happening, and they can communicate their sense of understanding to supporters. Frames become a theory of the situation. They produce not only a model of the current state of affairs but also suggest a paradigm for subsequent action. Frames name the elements of importance in the story and ignore those not thought essential by the framer. Frames provide the context in which to understand how the issue has been transpiring, what is currently emerging, and what needs to be done. Frames seek to tell a story, to weave together a coherent narrative, and to move beyond mere problem definition toward persuasion (van Hulst and Yanow 2016).

Framing cannot only affect how a policy issue is understood, but can also be instrumental in determining the later steps in policy analysis and design. For instance, in the first stage of analysis—understanding the issue—framing is used to set the stage for debate over a policy issue in three ways. In problem detection and initial framing, the framing will determine whether there is a problem to be addressed at all.

In the next step of information gathering, framing will determine the parameters of information gathering. Should it be narrow or widely cast? Should it be short-term and technical or long-term and strategic? At the problem definition stage, framing will serve to provide the agreed upon definition of the problem.

When the analysis moves on to the planning and decision making stage, framing will help determine the options considered by framing the range and nature of alternatives considered. Framing will determine the assessment of options by denoting the kind of expertise that will be required by each option. The framing of uncertainty about unintended consequences will assist in determining the selection of alternatives.

Managing the policy solution will also be affected by framing. How the players frame the policy options will affect implementation. Monitoring will be affected by the framing of the monitoring requirements, and, finally, evaluation will be affected by the framing of the goals of the policy or program implemented as a solution (Dewulf 2013).

How an issue is framed can shift over time as new dimensions of the issue are explored. For instance, the American public's attitude toward the death penalty shifted in the wake of the rise of legal clinics and innocence projects that reframed the death penalty in terms of morality, efficacy, and fairness (Baumgartner, De Boef and Boydstun 2008). This process of reframing is often the work of policy advocates who seek to see the current status of a policy issue shift to one that they prefer. For instance, in the area of food policy, reframing efforts have included attention to labeling genetically modified organisms (GMOs) in food, requiring taxes on sweetened beverages, and allowing raw milk into the food supply chain (Rahn, Gollust and Tang 2017).

Reframing plays a critical role in understanding how opinion, whether by the public or elected officials and their staffs, changes regarding a policy issue. Of course, political attitudes can shift as a result of elections thought to be a mandate on a particular policy or issue, however, opinions can and do change without an electoral change. Periods of political stability, in which policy monopolies dominate, are maintained by the policy monopolists tying their policy to strong images and symbols such as progress, national identity, and economic growth. This positive framing helps to underpin the monopoly position as well as serving to scare off potential opposition to the policy. However, if the opposition can redefine the policy issue to their advantage through providing alternative framing or counter-framing, they can attract previously uninvolved citizens into the debate and perhaps shift the views of those already active in the policy debate. When such reframing occurs, attitudes can shift, and policy changes can be rapid (Baumgartner and Jones 1993).

The effectiveness of framing can be impacted by whether the public is familiar with the issue. A new issue that is unfamiliar and/or complicated is more likely to be influenced by framing than an issue that is long in the public eye, even if still unresolved (Lee and Chang 2010). The effectiveness of framing is also influenced by partisan identity. Political polarization leads voters to rely more on their own partisan identity when making political decisions, leading to less critical thinking and evaluation of facts. To overcome partisan divides, policy communication seeks to shape public opinion through framing. Partisan identities are also social identities. And while a partisan identity might lead someone, for instance, to oppose spending on education, the social identity of being a parent might overcome that initial decision. Shifting the lens from partisan identity to another social identity through framing has been shown by research to be more effective in changing a voter's policy position (Diamond 2020).

In the behavioral decision-making literature, one of the most common types of framing discussed is gain-loss framing which has been used in many areas of communication including health communication to increase organ donation, anti-smoking campaigns, framing political statements to focus attention on equal protection and free speech, marketing campaigns to encourage travel, science communication to raise earthquake preparedness, and climate change debates to increase willingness to sacrifice to solve the problem. Gain-loss framing was elaborated by Kahneman and Tverskey as part of prospect theory. The theory seeks to explain how people make decisions given probabilistic information under conditions of uncertainty.

This theory places emphasis on what is to be gained or lost, explaining that quite different decisions could be made based upon the same information whether it was framed as a loss or a gain. Gain-framed messages highlight positive outcomes while loss-framed messages emphasize negative outcomes. Under this theory there are four possible frames. The first two are gain-framed and they include gaining a positive outcome and avoiding a negative outcome. The last two are loss-framed and they include experiencing a negative outcome and missing out on a positive outcome.

Gain-framed messages have been alleged to be more successful in promoting risk-adverse behaviors while loss-framed messages have been purported to be more successful in promoting risk-seeking behaviors in which greater risk is accepted with the hope of ending up with a better outcome. However, studies have shown that the differences between gain-framing and loss-framing over a wide array of issues are minimal, suggesting there may be other intervening variables at work (Nabi et al. 2020).

Framing is also identified by the debate over emphasis versus equivalence framing, which centers on whether competing frames should contain different conceptual content or not. Equivalence frames present different but logically equivalent content, promoting consideration of the same information. For instance, equivalence framing a debate about crime policy may use the gain-frame of keeping 75 percent of criminals off the streets versus the loss-frame of keeping 25 percent of offenders on the street. The question asked by scholars is which sort of framing is more persuasive?

Equivalence framing is rooted in psychology and looks at what affects an individual's perception. Emphasis framing, on the other hand, promotes consideration of different issues. For instance, in covering elections, one frame might provide information about a candidate's policy positions, while another frame might only emphasize the candidate's polling percentages. Older research on political framing typically assumed equivalence framing; more recent research on political framing, however, shows a far more prevalent use of emphasis framing (Brugman and Burgers 2018). In the policy debate over nuclear power, for instance, emphasis framing can communicate a negative frame of safety by calling attention to accidents or a positive frame that draws attention to the lack of carbon dioxide emitted by nuclear power plants. These competing frames (frame and counter-frame) play out in the policy debate (Kobayashi 2020).

As just noted, political framing also includes the concept of counter-framing. As one side seeks to emphasize certain aspects of an issue, their political opponents will try to recast or alter the picture with a counter-frame. This can be seen clearly in the framing of issues like abortion where one side depicts the issue as one of privacy and a woman's right to control her own body while the other political side seeks to frame the issue as one of morality and the rights of the unborn. Likewise, in the issue of gun control, one side frames the issue as a Second Amendment right while the other side frames it as controlling crime or gun violence. Events can intervene in the framing and counter-framing of policy issues. In the gun control issue, for instance, a focusing event such as a mass shooting can raise the saliency of the issue and give both sides the opportunity to pitch their preferred frame at the public again.

Central to framing and counter-framing is the desire of political opponents to point out the flaw in the opponent's messaging (Callaghan and Schnell 2005).

Actors in Policy Framing

The media is a key actor in political issue framing. The presentation of an issue from one perspective while excluding other perspectives can determine how the public will perceive an issue. For instance, Nelson, Clawson, and Oxley's study on how the public perceived a Ku Klux Klan rally concluded that results shifted based upon whether it was framed in terms of freedom of speech or disruption of public order. They found that those participants in the experiment that were exposed to the freedom of speech frame were more tolerant of the KKK rally than those who were exposed to the disruption of public order framing (Nelson, Clawson and Oxley 1997).

The media can also frame issues using a more generalized approach rather than an issue specific frame. For instance, media framing in the wake of the Three Mile Island and Chernobyl nuclear power plant accidents used generalized imagery of radiation detectors and mushroom clouds to produce a generalized frame of public fear and current or future loss. Similarly, the media can frame an issue to emphasize potential gain, such as media coverage of health care reform as a gain for coverage equity as opposed to a loss-based frame that might emphasize the loss of personal freedom (Boydstun and Glazier n.d.).

The role of experts in framing policy issues is also of importance. Individuals may accept a frame put forward by an expert if she or he is thought to be credible, trustworthy, or holding a certain level of expertise. Research has also shown that whether an individual accepts a frame being put forth by an expert is also dependent on the individual's predisposition. Research has shown that in the early stages of opinion formation, individuals may rely more heavily on their own prior values, beliefs, and opinions than on expert information. They may resist acceptance of an expert frame that conflicts with their own predisposition but readily accept an expert frame that is not in conflict with their predisposition (Lachapelle, Montpetit and Gauvin 2014).

Framing is undertaken by politicians and political organizations to sway public perception. Politicians are in almost continuous interaction with the media, lobbyists, interest groups, and businesses, all of which are attempting to frame issues in advantageous ways to sell their position to political decision makers. Politicians themselves are almost constantly trying to frame the issues they deal with in ways that will earn them credit from voters and the public (Sheffer and Loewen 2018).

George W. Bush's War on Iraq, for instance, was framed by his administration in several different ways in an attempt to win public support. First, the administration argued that the war was necessary for U.S. homeland security. When that argument did not seem to gain public support, then the administration framed it in terms of weapons of mass destruction. When the evidence showed there were no weapons of mass destruction in Saddam Hussein's possession, the frame shifted again. Finally, the Bush administration simply made the argument that a world without Saddam Hussein would be a better world (Lee and Chang 2010).

Research on framing by policy makers shows that they have a set of concerns different from other framers (like the media) because they need to explain their positions to voters and continue to appear credible to colleagues. These concerns may affect their strategies for framing and reframing (Mucciaroni, Ferraiolo and Rubado 2019).

Framing Transitions in the Climate Debate

Changing the language used to talk about what is happening to our climate is nothing new. From the beginning of human understanding of the phenomenon, global warming was the term used, in large part to underscore the process of greenhouse gases (GHGs) warming the planet. That discovery dates to 1896, when Svante Arrhenius published a paper linking the burning of fossil fuels to increases in average carbon dioxide levels in the atmosphere. He found that atmospheres with more carbon dioxide would retain more heat. This led him to hypothesize the greenhouse effect and global warming. As late as the 1950s, this warming was not generally regarded by scientists as a problem but that would change when Charles Keeling's data from Mauna Loa, Hawaii, began to show ever increasing amounts of carbon dioxide in the atmosphere year after year (Rahm 2010).

In 1988, Dr. James Hansen of NASA testified before a Congressional committee that the problem was no longer theoretical. Hansen told members of Congress that he had observed data that indicated warming had already begun. This testimony was heavily covered by the press and served to accelerate the general public's awareness of the issue of global warming. With the release of *The End of Nature* in 1989, the public became even more aware of the global warming issue (McKibben

1989). This widely read book popularized the thesis that industrialized societies had release enough carbon dioxide and other warming GHGs into the atmosphere to make the atmosphere no longer nature's creation. McKibben argued that with these releases the content of the atmosphere was now determined by humanity rather than nature. He further argued that consequently humankind could expect a warmer world with changes to weather patterns.

By the late 1980s and early 1990s, a growing number of scientists and government agencies including the National Aeronautics and Space Administration (NASA) and the Environmental Protection Agency (EPA) began to warn that the world should take action to reduce greenhouse gas emissions because of the untoward effects such warming would have. These scientists and agencies warned of melting of glaciers, rising sea-levels, more severe weather events, enhanced forest fires, species extinction, and ocean acidification.

As it became clear that global warming was more than just warming, the use of the phrase climate change became more common, and the frame shifted. Indeed, the creation of the Intergovernmental Panel on Climate Change (IPCC) and the release of its First Assessment Report in 1990 fueled concern over the negative impacts of this warming in terms of specific shifts in the climate system, thus leading to the United Nations Framework Convention on Climate Change (UNFCCC). While only a framework treaty, the UNFCCC began the process of international negotiation of a binding treaty to bring down atmospheric levels of greenhouse gases (Rahm 2010).

Some of the earliest counter-framing used in the climate change debate was associated with climate denialism. This narrative was heavily spun by the fossil fuel industry, conservatives, and conservative think tanks seeking to delay any policy response to climate change. The initial "merchants of doubt" as they came to be called, dated to the 1990s and they used two primary anti-science frames to convince policy makers that the issue did not need political intervention.

The first of their tools was to frame climate science as inaccurate. To accomplish this, they argued uncertainty in the scientific consensus surrounding global warming, often arguing that no warming was taking place or that, if it was, such warming would be beneficial by lengthening the growing season, for instance, or opening the Arctic for oil and gas drilling. They specifically pointed to the transport benefits associated with opening the legendary Northwest Passage, the sea route long sought from the Atlantic to Pacific Ocean through the Arctic.

Later, as the scientific consensus around climate change strengthened, the denialists shifted their framing to include economic arguments.

This loss-framing attempted to suggest that the costs of policy intervention were much too high to justify any policy action, especially in terms of the cost to energy consumers (Cann and Raymond 2018).

Early in the debate on climate policy, the frame desired by activists was sharply focused on mitigation, or reduction of GHGs. There was little or no discussion of adaptation. Activists feared that to focus on learning to live in a warming world would turn attention away from mitigation efforts. Over time, however, as physical evidence of the negative outcomes of warming emerged, it became necessary to reframe the problem not just as one of mitigation but mitigation and adaptation. With climate change, the issue frame shifted as evidence mounted of the harmful impact that climate change was having.

This understanding shifted the frame from a singular effort of reducing GHGs to a broader one of both mitigation and adaptation. Much of this reframing came in the form of emphasis on the economic costs of inaction (a loss-frame), and the economic benefits (a gain-frame) of a switch to renewable energy, in particular, job creation (Diamond 2020). Floods and droughts were early physical impacts from climate change that would force a reframing to enable meaningful policy actions for communities to respond to these events.

A curious phenomenon also arose in that communities could respond to embedded policy problems related to the impacts of climate change such as drought and floods without a mention of climate change at all. Despite that fact that floods and droughts were made more severe by climate change, political opposition to recognition of climate change (climate denial) compelled many local policy makers to avoid mention of causes and simply address the disaster at hand (Hurlbert and Gupta 2016).

During the 1990s and early 2000s public awareness of climate change increased. Part of the rising awareness was due to the negotiation of the Kyoto Protocol in 1997, the first binding treaty to result from the Climate Secretariat created by the United Nations Framework Convention on Climate Change. This treaty was very controversial in the United States, despite the support of the Clinton administration, including then Vice President Al Gore, a known environmentalist. The loss-framing of climate change as too costly to do anything about succeeded in turning the Congress against supporting the Kyoto Protocol. President Clinton never submitted the treaty to the Senate for ratification and President Bush, who fully accepted the loss-framing, did not prioritize any policy actions to seriously deal with mitigation (Rahm 2010).

In 2001, Frank Luntz, a Republican strategist and political consultant, had written a memo on environmental talking points to Republican politicians advising them to stop using the phrase global warming because it had catastrophic connotations and to switch to climate change because it was more neutral and less emotional (Zak 2019). But later, Luntz changed his view saying that the public reacts to both terms about equally. Scientists, though, had come to prefer the phrase *climate change* as it represents the wider array of phenomena associated with warming—such as ocean acidification—that are not as narrow as temperature increases. The framing of the issue as climate change allowed the understanding that the world was not just getting warmer, it was getting more extreme with characteristics of global climate disruption (Schumacher-Matos 2011).

After winning the popular vote but conceding the highly contested presidential race in 2000 to George W. Bush due to a Supreme Court ruling, Al Gore turned again to climate activism. Gore had previously established his ecological positions, having published *Earth in the Balance* in 1992 (Gore 1992). The book became a best seller. Gore had long been involved in environmentalism. As a freshman member of Congress in 1976, Gore held hearings on climate change. After his presidential loss, Gore created an educational PowerPoint presentation on climate change. He presented it many thousands of times.

A documentary film maker, Davis Guggenheim, used it as the basis of his Academy Award–winning 2006 movie *An Inconvenient Truth* (Gore 2006). Gore also published a book under the name *An Inconvenient Truth: The Planetary Emergency of Global Warming and What We Can Do About It* which framed the climate issue as an emergency (Gore 2006). The following year, Gore published *An Inconvenient Truth: The Crisis of Global Warming* aimed at the youth population (Gore 2007a). In his 2007 Nobel Peace Prize Lecture, Al Gore referred to the climate crisis and to the planetary emergency (Gore 2007b). Gore shared the 2007 prize with the IPCC. The success of the film *An Inconvenient Truth* led Gore to create the Climate Reality Project to train local leaders to inform their communities about the climate crisis and to organize action to fight it (Morrison 2017). Gore had been an early adopter of the crisis or emergency frame.

Climate change continued to be an international issue of extreme importance as the final years of the Kyoto Protocol began. The Kyoto Protocol was originally expected to expire in 2012 but was extended through 2020 due to the difficulty of negotiating a successor treaty. A breakthrough in diplomacy led to a new path forward that included all the nations of the world in the successor treaty. The Kyoto Protocol had allowed developing countries to avoid binding reductions of GHGs, a stipulation that had reinforced the loss-framing of the treaty by American opponents who argued that demanding developed nations take action, without requiring developing nations to do anything, would cause financial harm to the United States. By adopting a new approach of each nation determining its own reduction goals, called Nationally Determined Contributions or NDCs, a roadblock to the successor treaty was bypassed. In 2015, the Paris Agreement was agreed to by 195 nations. This agreement went far in adopting the more urgent framing of the issue by setting the treaty's goal to keeping average temperature rise to "well below" 2 degrees Celsius above pre-industrial levels and trying to limit the rise to 1.5 degrees Celsius (Rahm 2019).

Beginning in the late 2010s, the more urgent framing of climate change emerged from a variety of places. In October of 2018, the IPCC issued a report indicating that there were only 12 years left for global warming to be limited to 1.5 degrees Celsius and that urgent actions needed to be undertaken to accomplish that goal. The report also stressed that the difference between a world with 1.5 degrees of warming and 2 degrees of warming would be vastly different. The additional half degree of warming would mean much worse conditions. The report also upgraded the scientists' sense of certainty (Watts 2018). In late 2018, the U.S. House of Representatives established the House Select Committee on the Climate Crisis. The original House committee on the issue had been formed in 2007 was called the Select Committee on Energy Independence and Global Warming, however, that committee was abolished by Republicans when they took control of the House in 2011 (Meyer 2018). Also in 2018, UN Secretary-General António Guterres used the term crisis in a speech he gave at the United Nations (Rigby 2020).

Much of the media shifted its stance on language, as well. In May of 2019, *The Guardian* adopted style changes that preferred the use of such wording as climate emergency, crisis, or breakdown or global heating over climate change, although the latter was not banned (Carrington 2019). In June of 2019, the Spanish news agency EFE also announced it would use the phrase *crisis climática* (climate crisis) and then Spanish language Noticias Telemundo announced a shift to climate emergency. The urgent phrases were also appearing in CNN headlines and the TED Radio Hour (Yoder 2019). The use of the new crisis frame language spread worldwide in 2019, including to the *Hindustan Times* (Editorial 2019).

Not all media adopted the language change. For instance, the CBC news said that the use of *crisis* suggested advocacy and urged caution among public broadcasters to maintain neutrality. However, critics argued that given the existential threat that climate change poses, reporting on this issue should be something more than neutral and passive. Indeed, they argued that neutrality favors the business-as-usual approach, which itself is a bias. By being neutral, they argued, the media brings unwitting support to those industries and individuals that profit from contributing to climate change (Takach 2019).

In the fall of 2019, as the push for formal declarations of climate emergencies increased in North America and western Europe, parliaments in the United Kingdom, Ireland, France, and Canada passed resolutions of climate emergencies (Hulme 2019). In June of 2019, New York City declared a climate emergency, making it the largest city to do so. By so doing, it joined London, Sydney, and 722 localities in 15 countries that had also declared a climate emergency. Groups like the Sunrise Movement and Extinction Rebellion have made such declarations centerpieces of their platforms (Barnard 2019).

The rising role of populist climate movements was evident by 2019. Beginning in May of 2019, Swedish activist Greta Thunberg gave up the use of the phrase *climate change* in favor of *climate breakdown* or *climate emergency* (Zak 2019). Thunberg began her climate activism with a school strike to draw attention to the climate emergency and the fact that politicians worldwide were taking insufficient action to address the crisis. The group Fridays for Future was formed in 2018 after Thunberg and other activists sat in front of the Swedish Parliament every school day for three weeks. Her calls for strikes grew and in September of 2019 an estimated four million people participated in global climate strikes (Marchese 2020).

The Sunrise Movement, another of the rising populist climate movements, is a youth movement focused on stopping the climate crisis. Sunrise aspires to other goals as well. As supporters of the Green New Deal, their strategy to curb greenhouse gases includes the creation of millions of green jobs. The Green New Deal was a Congressional resolution introduced by Representative Alexandria Ocasio-Cortez and Senator Ed Markey during the 119th Congress in 2019 to move the country quickly to renewable energy. The Sunrise Movement adopted a strategy of recruiting millions of young people "to make climate change an urgent priority across America, end the corrupting influence of fossil fuel executives on our politics, and elect leaders who stand up for the health and wellbeing of all people" (Sunrise Movement n.d.).

The movement was started in April 2017 with the initial goal of making climate change matter in the midterm 2018 elections. Sunrise rented space from the Sierra Club in Washington, D.C., and received a grant of $50,000 from the Sierra Club Foundation (Matthews, Bowlin and Hulac 2018). Sunrise was charted as a 501c(4) (Sandoval 2018). The movement, organized into hubs spread across the U.S., participates in political activities (Sunrise Movement n.d.).

Extinction Rebellion, another of the populist climate movements, is a global environmental group working to get governments to act on the climate and ecological emergency. It emphasizes both the climate crisis and the Sixth Great Extinction. It rebels against the systems that got the planet into the environmental crises it now faces. Its home page opens with a statement: "THIS IS AN EMERGENCY" (Extinction Rebellion n.d.). The organization was begun in the UK in 2018 with the publication of an open letter published in *The Guardian* and signed by 94 academics demanding that the "government to take robust and emergency action in respect of the worsening ecological crisis" (Green and Cato 2018).

In May of 2019, Al Gore's Climate Reality Project put out a call to all major news outlets to call climate change a climate crisis, arguing that it was the responsible thing to do (climaterealityproject.org 2019). In June of 2019 Public Citizen sent a letter to news organizations urging them to call climate change a crisis. The letter was signed by several progressive groups, including Sierra Club, Greenpeace, and the Sunrise Movement (Yoder 2019).

The extent of this reframing effort resulted in the *Oxford English Dictionary* naming *climate emergency* as the "word" of the year for 2019, saying the use of the phrase had increased by 10,796 percent. The dictionary said the increase in use was reflective not only of an awareness of the issue, but also of the conscious push toward language of urgency and immediacy (Zhou 2019).

In a November 2019 statement published in *Bioscience* in January 2020, a worldwide coalition of 11,000 scientist signatories stated the call for urgent action. The scientists asserted that they had a "moral obligation to clearly warn humanity of any catastrophic threat" and that "clearly and unequivocally the planet Earth is facing a climate emergency" (Ripple et al. 2020, 1). In that same month, scientists published an article in *Nature* on climate tipping points that they think underscore calls for urgent action. The tipping points cited included: Amazon rainforest frequent droughts, reduction in Arctic Sea ice, slowdown of the Atlantic circulation, fires and pests changing the boreal forests, large-scale die-offs of coral reefs, accelerating ice loss of the Greenland ice sheet, thawing of the permafrost, accelerating ice loss of the West Antarctic Ice Sheet, and accelerating ice loss in the East Antarctic ice sheet. The article referenced the IPCC Special Report of 2018 that suggested that individual tipping points could be triggered with as little as

1–2 degrees Celsius temperature rise and that the globe had already seen increases of nearly 1 degree. It ended with a chilling call for action, saying "The stability and resilience of our planet is in peril. International action—not just words—must reflect this" (Lenton et al. 2019, 595).

With the election of Joe Biden, the climate crisis received renewed political momentum. On the day he announced his first cabinet appointments, the president-elect also announced the creation of a new cabinet-level post, that of Climate Envoy. John Kerry, the Secretary of State in the Obama administration, was named to the post. Kerry had helped negotiate the Paris Agreement for the Obama administration. With Kerry as a sitting member of Biden's National Security Council, the issue of climate change was elevated to the highest level of government. Kerry, in a statement after the announcement was made, declared he intended to treat the climate crisis as an urgent national security threat. Biden, as well, promised to work with Congress to address the climate emergency (Friedman 2020a).

While President Trump had withdrawn the U.S. from the Paris Agreement on November 4, 2020, Biden indicated that he would, on his first day in office, return the U.S. to the Paris Agreement (Grandoni and Ellerbeck 2020). He did just that on his first day in office. He also signaled the importance of the climate crisis by halting the Keystone XL pipeline, placing a moratorium on oil and gas drilling on federal lands, initiating a process to invest in low-income and minority communities that have historically borne the burden of high levels of pollution, and demanding that climate change become a priority for almost all federal agencies (Tharoor 2021). Biden also named Gina McCarthy, the EPA Administrator in the Obama administration, to head the White House Office of Climate Policy thus establishing both an international and domestic leader on the issue (Friedman 2020b).

Biden's initial attempt to push forward his Build Back Better Framework, which contained many provisions linked to the climate crisis frame, was thwarted due to lack of sufficient votes in the Senate. However, by pushing the framework, the Biden administration seemed to accept the crisis framing. Critics, especially from the left, argued that other actions undertaken by the administration such as opening public lands to oil and gas drilling suggested such a crisis commitment was not forthcoming.

Historical Outcomes of Crisis Framing

Perhaps history can tell us something of crisis or emergency framing. Emergency framing of the nuclear war issue occurred in the early

1980s in the United States, Western Europe, and Japan. In 1980, Helen Caldicott, a leading anti–nuclear war advocate, insisted that the time frame to save the world from the threat of nuclear war was six months. She was convinced that the election of Ronald Reagan could lead to nuclear war and the subsequent destruction of humankind. By 1983, the theory of "nuclear winter" circulated widely, once again suggesting the complete annihilation of humankind.

Nuclear winter was a theory that hypothesized that an exchange of weapons would result in many fires that would emit vast amounts of smoke into the atmosphere. These smoke particles, it was argued, could block significant amounts of sunlight from reaching the surface of the earth, thus vastly cooling the planet. Such an event, if it were to occur, would cause a cascade of secondary crises, such as failure of the food production system. Atmospheric scientist Carl Sagan used this theory to call for immediate drastic cuts in weapons.

During the 1980s, millions of people marched in the public protests that arose over nuclear weapons. A widely watched television movie, *The Day After*, vividly showed the aftermath of a nuclear exchange on the American midwestern population, popularizing the emergency framing of the issue. But by the end of the Cold War, the worldwide nuclear weapons opposition movement simply faded out. And yet this was not because the danger had passed. The *Bulletin of the Atomic Scientists* doomsday clock remained close to midnight (Hodder and Martin 2009).

Framing this issue as a crisis, or an emergency, may have had some effect on policy solutions if the widespread public outcry pushed political leaders to negotiate arms reductions. Talks for the first Strategic Arms Reduction Treaty (START) began in 1982 between the Reagan administration and the Soviet Union. By 1987, as the Soviet Union was nearing its coming demise, talks resumed at the Reykjavik summit between Ronald Reagan and Mikhail Gorbachev. But it would not be until after the fall of the Soviet Union in 1991 that START would be signed by George H.W. Bush and Mikhail Gorbachev (NTI: Building a Safer World 2011).

Negotiations for reductions in nuclear weapons had begun long before the issues gained popular crisis framing. Initial talks for what would become the Strategic Arms Limitation Treaties (SALT I and II) were suggested by the Johnson administration in 1964 and went into effect in 1969; SALT I and II would then go into effect in 1972 and 1979, respectively. These treaties clearly cannot be the result of the framing of the 1980s, as they predate them. And diplomacy following the 1979 treaty neatly transitioned to the START negotiations (Britannica 2020).

Crisis framing of nuclear weapons and war obviously did not result in the elimination of nuclear weapons. Russia and the United States in 2020 together possessed more than 90 percent of the world's nuclear weapons (Ploughshares Fund 2020). In 2020, the world's nuclear nations had more than 13,500 weapons. The other nations with nuclear weapons include the UK, France, Israel, Pakistan, India, China, and North Korea. Nations like Iran aspire to have such weapons (ArmsControlAssociation 2020). The nuclear issue then shows that framing an issue as a crisis does not necessarily mean that political action will ensue.

History perhaps can also show evidence of successful crisis framing. Dealing with the hole in the ozone layer is one such instance. In 1984, British scientists working in Halley Bay Station in Antarctica published a study reporting their measurements of ozone in the region. They had found that each spring the amount of ozone in the Southern Hemisphere had dropped sharply since the 1970s and that this corresponded to the level of chlorofluorocarbons (CFCs) measured in the atmosphere. They had no scientific proof but nevertheless they organized a press conference to popularize their finding. The discovery of the ozone "hole" was major news, attracting a great deal of public attention and provoking a sense of psychological dread in the public. Fears of skin cancer, cataracts, and crop damage created a movement on the part of the public worldwide that demanded action.

Such action was forthcoming even before the scientific evidence surfaced to prove the relationship of CFCs to ozone depletion. The Vienna Conference was adopted in 1985 and it led to the Montreal Protocol on Substances that Deplete the Ozone Layer by 1987. This rapid progress was owed to many factors but one of them most certainly was the urgent framing of the issue. However, it is important to emphasize that eliminating a refrigerant manufactured by a relatively small number of companies based exclusively in the West is a scenario very different from climate change.

To address the problem of ozone depletion, alternatives to CFCs had to be developed and DuPont rapidly did so. It is also important to note that DuPont knew that its more expensive alternatives would not gain market share without an international agreement on ozone depleting substances and so DuPont supported rather than tried to block the treaty. Climate change is a more extensive problem that will require an economic transition far greater than eliminating ozone depleting substances from the economy (Rahm 2019).

Potential Negative Consequences of Emergency Framing

When a policy issue like climate change receives little genuine response by policy makers, advocates of policy action may turn to branding the policy issue an emergency. Emergency framing, they suggest, will provide the necessary energy to overcome policy inertia. This scenario played out with the climate debate as its depiction moved to climate crisis. But this framing is not without potential negative outcomes.

If scientists and advocates are thought to exaggerate the dangers of changes to the climate, opponents may dismiss the issue entirely by calling them alarmists. Since there are already so many politicians and members of the public that say they doubt the scientific consensus on climate change, they may easily react negatively to alarming messages that replace the normally careful tone of scientists (Risbey 2008).

While trying to motivate human behavior through fear may be successful especially if it resonates with personal experience, since climate change is typically seen as a distant and slow-moving issue, resorting to fear as a motivator may not work. Fear also may trigger the response that the issue is overwhelming and can lead to denial or inaction (Hodder and Martin 2009). Marketers use fear as often as they can to motivate action. But the key is not just to scare, the key is to scare and then offer a specific solution that can be adopted (Wilson 2009). This is very tough to do with the issue of climate change, so the likelihood that fear will motivate behavioral change around this issue is not great.

Adopting a crisis frame may arouse fear, but without proffering a solution, denial or inaction are likely to result. This is one reason why so many people react to climate change by simply feeling overwhelmed. Adopting this "doomism" frame is a recent twist in climate denial counter-frame. The use of this psychological tool by industries responsible for most of the GHG emissions worldwide results in paralysis by at least some members of the public. If the public falls into political immobility because of the psychological acceptance that the planet is doomed, the forces that seek to gain from political inaction will champion (Mann 2021). Scholarship shows that countering the doomist framing of climate change requires climate change activists to find ways to incorporate hope into their messages, while at the same time not evading the scientific facts (Kelsey 2020).

Adopting an emergency frame to climate change prioritizes climate change above other important issues and could encourage competition among activists. There are many important public policy issues around which activists mobilize and emphasizing climate as the top priority may cause resentment in some. Why should we declare a climate emergency but not a food emergency in a world where many are food insecure? Or a water emergency in a world where many experience water scarcity? Or a violence emergency in a world where many are unsafe? And the list can go on as we review the wide range of public policy issues of concern to many (Hodder and Martin 2009). Climate change activists need to approach this issue with caution. One way to avoid the negative outcomes could include linking other issues such as food scarcity and water shortages to climate change, thus showing those concerned with other issues that they are in fact linked or made worse by climate change.

By adopting an emergency frame, slower solutions may be abandoned while riskier solutions, like geoengineering, may be advanced. Geoengineering was introduced by Nobel Prize winner Paul Crutzen in 2006 when he suggested seeding the upper atmosphere with sulfur to reflect sunlight back into space. Geoengineering approaches fall into two categories: Solar Radiation Management (SRM) and Carbon Dioxide Removal (CDR). Crutzen's approach, dispersing stratospheric aerosols, became one of the key ideas in SRM. Critics of SRM argue that while it could lower temperatures, it would have no effect on the cause of global warming, the increasing greenhouse gasses in the atmosphere, and therefore would be ineffective in dealing with other effects like ocean acidification.

Of the two approaches, CDR is preferable, as it does get to the cause, but it is slow and expensive. SRM is both affordable and readily doable with current technology, so it is seen as the most likely candidate for geoengineering experiments. Several entrepreneurs and startup companies are actively pursuing CDR technologies however they are at an early stage of development and not yet commercially viable on a large scale (Rahm 2019).

Widely deploying such a technology would depend on determining who would be declaring such an emergency. In the case of geoengineering, it would likely be a select group of specialized experts such as climate scientists or commercial firms with the mission to develop and deploy such technologies. But this has led critics to argue that this would erode democracy and lead to decision making by technocracy or a small number of individuals not fully accountable to the public (Horton 2015).

Framing climate change as an emergency may suggest that only action from powerful centralized governments can result in success and that individual actions are less important (Horton 2015). This ties to the doomism frame suggested earlier. If people become convinced that individual action will be ineffective, they may fail to take any action at all. On the other hand, part of the recent efforts of the fossil fuel industry has been to deflect attention from their overwhelming contribution to the climate crisis by pushing the notion of individual responsibility for climate change. The fact is that while individual effort is important, without the adoption of widespread reductions by the industrial sector, the battle over climate change will be lost (Mann 2021).

Emergency framing may also concentrate efforts on achieving the single goal of net-zero emissions by a specific date and largely focusing on transitioning away from fossil fuels to do it. Some suggest the Green New Deal does just this. In doing so it may actually deter us from taking on a wider array of solutions that will collectively lead to solving the problem (Hulme 2019).

These other solutions have been well documented and include such seemingly unrelated matters as educating women and girls, reforming agricultural practices, ending deforestation, reforestation, moving to a plant-based diet, ending food waste, and dealing with leaking refrigerants which are powerful greenhouse gases, amongst others (Hawken 2017). Indeed, a 2020 study reported in *Science Magazine* analyzed the various paths to achievement of the Paris Agreement's goal of limiting temperature increases to 1.5 or 2 degrees Celsius above preindustrial levels.

That study showed that even if fossil fuel emissions were stopped immediately, business as usual in the agricultural sector would prevent achievement of the 1.5 degree goal, and, by the end of the century, threaten the achievement of the 2 degree target. The global food system must change if the Paris Agreement goals are to be met (Clark et al. 2020). If the emergency framing so fully focuses on achieving the one single goal of net-zero emissions from the power sector, this alone could block the inclusion of the agricultural sector in the solution set.

The world is in the midst of the fourth industrial revolution with rapidly emerging technologies in the fields of artificial intelligence, material science, digital communication, and genomics—all with the potential to vastly change the way humans live and work. Equally important is the fact that the world is experiencing other important shifts including population growth, increasing economic inequality, food and water scarcity, and new doubts about the durability of our institutions. Dealing with climate change over the next decades will have to take these changes into account as well, for failing to do so could make the world less hospitable to many even if it is powered by net-zero carbon. Assuring a positive future might better be achieved by concentrating on the wider array of UN Sustainable Development Goals, all 17 of them, rather than one metric (Hulme 2019).

Emergency framing seems to create a vision of a single massive problem that can be solved only by international treaties and strong central governments. But others suggest that solving climate change is better approached as a series of discrete, manageable problems that can be beaten. By working on each simultaneously, each part of the solution makes our societies stronger and healthier. The problem is not one massive thing but rather many things that can be addressed by actors other than strong central governments—cities, states, businesses, and citizens (Bloomberg and Pope 2017).

Since the crisis framing of climate change has already occurred, climate change activists must be vigilant of the potential downside of that framing and seek to address issues as they may arise. Claims of exaggeration or alarmism need to be responded to with firm statements of alarm tied directly to the science. Attempts to change behavior by fear need to be accompanied by solutions as to what can be done. Acknowledgment of the importance of many policy issues needs to reassure advocates that climate change will not drive other issues into obscurity. Risky solutions like geoengineering need to be subject to democratic governance before being deployed. Adoption of a single solution of net-zero emissions by a prescribed date through widespread adoption of green energy technologies must be understood to be only part of the solution—not the whole solution. A top-down-only approach to solutions should not be adopted.

Conclusion

The discussion about climate has shifted over time. The issue of climate was first framed as the scientific phenomenon of greenhouse gases causing global warming. This early framing did not present the issue as a problem, rather early framing argued that such warming might even be advantageous by allowing longer growing seasons. As scientific studies continued to look into the consequences of ever increasing levels of carbon dioxide in the atmosphere and the resultant warming, concerns were raised by advocates, government agencies, and scientists. The issue was reframed as a problem that needed attention.

As scientific knowledge of the potential results of global warming increased, scientists and advocates reframed the issue again by moving away from the phrase *global warming* to use of the term *climate change*. This shift was largely the result of greater understanding of the wider impacts of planetary warming which would include sea-level rise, ocean acidification, coral reef die-offs, loss of species, severe storms, droughts, and massive forest fires. As the many cascading effects of warming became clear, climate change seemed to be a better way to frame the issue because it called attention to these many impacts of warming.

Pushback to the problem frame emerged with a counter-frame of climate denialism. The initial response was to simply deny the science. As the evidence mounted, however, opponents of taking actions to address the consequences of a warming world also relied on a loss-frame that emphasized the costs of taking action. This loss-framing was countered by advocates insisting on a gain-frame in terms of green job creation and avoiding the worst costs from the disasters that were sure to occur.

Early in the climate dialog, advocates for policy action fought to frame the issue with only one policy option—mitigation of greenhouse gases. As the effects of climate change began to be observed, however, it became clear that regardless of mitigation efforts, a certain amount of warming was bound to occur. With this revelation, advocates were forced to reframe the policy options into both mitigation and adaptation.

As evidence mounted over time, and as the effects of global warming began to be more readily seen, advocates for policy action moved to reframe climate change to a climate crisis or emergency. This reframing was widespread among advocates and scientists by 2020. The urgency of the issue was made clear by several scientific reports which announced that time was very limited to take action to avoid the worst effects of a more than 2 degree rise in temperatures. Whether this crisis framing will result in sufficient action taken to address the changing planet remains to be seen.

CHAPTER 2

Participants in
the Climate Change Debate

Introduction

Participants in the debate include both domestic U.S. and international actors. For the domestic debate these actors include the federal government, state governments, and local governments. The U.S. court system plays a special role in the climate change debate in that much of the legislation and or executive actions taken either by the federal government, the states or localities end up being challenged in the courts. The U.S. domestic nonprofit sector plays a critical role in the policy debate as well and while many nonprofits are U.S.-based only, others are U.S.-based chapters of international non-governmental organizations (NGOs). Both are discussed below. Within the U.S. and internationally, the climate change debate of much of the 21st century has been driven by youth. As with all policy issues, the media play a large role in maintaining public awareness of the issue. Climate change is a politically charged issue and efforts to reduce emissions known by scientists to cause increasing temperatures draw opposition from many politicians, climate deniers, and industries fearing loss of profits if their activities are restricted.

Internationally, efforts to confront climate change largely rest with the United Nations (UN) System, its many conferences, the climate secretariat, and various other organizations formed or supported by the UN including the UN Environmental Programme (UNEP), the UN High-Level Political Forum on Sustainable Development, the Global Environment Facility, the Intergovernmental Panel on Climate Change (IPCC), international environmental organizations, and the international youth movement. Each of these is discussed below.

U.S. Government Participants

Many U.S. government actors participated in the climate change policy debates in the first decades of the 21st century. These included elected officials, appointed officials, as well as hired staff support. A key overview of these actors follows.

Federal Government

The important role of the federal government in policy making for climate change highlighted the prominence of a few positions with no major shifts from patterns of the 20th century. That is to say that presidents, members of Congress, and appointed officials serving in government agencies played a substantial role. A key feature of the specific role played by each of these actors depended heavily on which political party they aligned with. The political divide that erupted in the 1980s between Republicans and Democrats on issues associated with environment persisted and exacerbated into the 21st century. Republicans consistently opposed the expansion of environmental regulation and in large part moved closer to or joined the coalition of climate deniers.

Apart from George W. Bush, who eventually publicly accepted the scientific consensus on climate change but refused to act on it, the next elected Republican president, Donald Trump, denied the scientific consensus on climate change. For Democratic presidents, the situation was reversed. Obama and Biden both accepted the scientific consensus on climate change and tried, so some extent, to enact policies to address the issue. This dynamic was echoed in the Congress with Republicans opposing action on climate change and Democrats demanding action. The pattern was also revealed in the executive agencies because of the presential appointment of its leaders. No other factor better explains the first several federal government administrations in the 21st century (Rosenbaum 2020).

The growing division between the parties as the Republicans moved from a traditional pro-business, low tax, deregulatory stance to one that embraced the populism of Donald Trump and what came to be called Trumpism, shifted national politics in fundamental ways. While the early Trump administration did achieve part of the traditional Republican agenda of lowering taxes, rolling back environmental regulations, and reshaping the judiciary, the populist issues of immigration, trade disputes with foreign countries, and rejection of many international norms that had persisted since the end of World War II soon became dominant. In addition to these shifts, the flood of misinformation; rise of racial tensions; protests over police violence; and the economic, social, and health chaos caused by the pandemic profoundly upset federal politics (Dimock and Gramlich 2021).

State and Local Governments

The partisan divide seen in the federal government was often duplicated at the state and local level. With the rise of populism, state and local governments in many places became the place in which many of the tensions spilled into the streets becoming local issues. Populist ideas, falsehoods, and conspiracy theories drove many local political actors (Lerer and Epstein 2021).

At the state and local level, however, the problems associated with climate change were becoming more of a reality. As fires, droughts, floods, intense storms, heat waves, and the effects of sea-level rise increased, many states and localities began to grapple with the reality. Resilience and sustainability became new important goals for many states and localities. In the interest of creating new clean energy jobs, many states and localities actively promoted the growth of renewable energy. Often even red states and localities could do this by consciously not associating these impacts with the politically charged discussion of climate change (Ricketts et al. 2020). However, in several of the more conservative states, like Texas, state leaders used their positions to pass bills that would threaten businesses and others that sought such political action as fossil fuel divestment or boycotts (Hernandez 2022).

The Special Role of the Courts

The courts play a special role in environmental policy through their role in interpretation of the law. They are often asked by industry, environmental groups, and other parties to review rulemaking actions taken by government agencies to assure that they have complied with statuary guidelines. Often this review is undertaken due to judicial review provisions of the Administrative Procedure Act which allow persons wronged by agency actions to seek redress in the courts. If the court invalidates the agency's rule, the agency is forced to begin the rulemaking process anew. The courts also play a role in environmental policy due to citizen suits. Many of the nation's environmental laws have provisions that allow individuals or groups to sue anyone or any organization that they believe to be in violation of national environmental laws. Environmental groups often use these provisions to sue the Environmental Protection Agency (EPA) or other environmental regulatory agencies to meet deadlines specified in the statues (Rahm, U.S. Environmental Policy: Domestic and Global Perspectives 2019).

Many of the big decisions involving climate change policy have found their way into the courts. Key among them include *Massachusetts v. EPA* in which the U.S. Supreme Court ruled that EPA had the authority under the Clean Air Act to regulate carbon dioxide emissions; the Clean Power Plan, the set of rules that EPA intended to use to implement its climate agenda; and *West Virginia v. EPA*, an effort by coal states to block EPA's ability to regulate carbon dioxide emissions. The courts will likely continue to be big players in the climate debate as both advocates and opponents use the courts in their attempt to win battles.

U.S. Non-Governmental Actors

Environmental Organizations

There are many environmental organizations active in the issue of climate change. They include mainstream groups that are large national or international organizations that follow moderate policies and are primarily membership driven. Environmental groups vary in their strategies and tactics, however. Some are quite moderate in their policies while others are more militant.

There are far too many mainstream environmental organizations to discuss in detail but a sample of some of the large ones provides some insight into this group of activists. One mainstream group is Audubon, which focuses primarily on birds. Audubon considers climate change the greatest threat to birds and support protecting species through protection of habitat and promoting a net-zero carbon emissions goal by 2050. Their climate initiative strategy is to work with state and federal governments to protect birds from the effects of climate change primarily by taking what they call sensible measures to protect species and habitats (Audubon 2022).

The oldest and one of the largest mainstream environmental organizations is Sierra Club. It focuses on a wider array of issues than a more specialized group like Audubon. Sierra Club seeks to provide access for all to the environment and places an emphasis on the core value of anti-racism. Sierra Club promotes every individual's right to clean air, clean water, and a stable environment. They promote conservation and seek to transform society with an emphasis on social justice. They frequently file lawsuits to achieve the ends they seek. In terms of climate, they believe that the climate crisis is upon us and, because of the persistence of racism, seek to promote climate justice (Sierra Club 2022).

The Environmental Defense Fund is a U.S.-based nonprofit that advocates for a wide array of environmental issues, including climate change. It works worldwide to partner with problem solvers to bring solutions, focusing on evidence and data to determine where it can have the largest impacts (Environmental Defense Fund 2022). Another mainstream environmental organization is the Nature Conservancy, which works in many countries focusing on conservation and biodiversity loss. Climate change is one of the areas addressed through partnerships with indigenous peoples and local communities (Nature Conservancy 2022). While there are other mainstream groups, this sample provides an overview of goals, strategies, and tactics.

Some domestic environmental groups are more extreme and have found themselves labeled eco-terrorists. Prominent among them is the Earth Liberation Front (ELF), which operates as anonymous autonomous cells or groups with the mission to inflict economic or physical damage on organizations that it believes harm animals or the environment. The ELF is thought to have merged with, or grown out of another organization, the Animal Liberation Front (ALF), which is known to have attacked university laboratories such as those at the University of Iowa in 2004, causing $450,000 worth of damages. The ELF is thought to have burned down a construction site of a 306-unit condominium complex, causing $22 million in damages. The group's primary tactic is arson (Senate Hearing 109–947 2005).

Earth First! is another of these organizations. It was formed in the 1970s by a group of deep ecologists that believed the established environmental community was ineffective and in collaboration with the corporate culture that it believes endangers the planet. Tactics include "grassroots and legal organizing to civil disobedience and monkey-wrenching." The group is known to have been responsible for violent attacks on public and private assets which it alleged were destroying the environment. Its slogan is "No compromise in defense of Mother Earth" (Earth First! Journal 2022).

The Rise of the Youth Movement in the U.S.

American youth, especially those in Generation Z (born after 1996), have grown up in an era of pandemic and with the climate crisis looming over their lives. These factors explain to a certain extent the overall mood of the youth movement, which is driven by rage, discontent, and a demand for equity and justice (Sengupta, Four Take Aways on the Youth Climate Movement 2022). In a 2021 study of climate anxiety in youth published in *The Lancet* that surveyed American youth aged 16–25, nearly 90 percent said they were "at least a little worried" about climate change. About 45 percent said they were "extremely" or "very worried" and nearly 20 percent said they were "extremely worried."

How this worry manifests itself is telling. Nearly 26 percent of American youth claim this worry impacts their functioning. More than half use words such as "sad," "anxious," and "afraid" to describe their feelings; 68 percent agree with the statement "the future is frightening" and 78 percent agree with the statement "people have failed to take care of the planet." These general feelings have direct impacts on youth's opinions about government. More than half believe that "the government dismisses people's distress," 62 percent think "the government is lying about the effectiveness of the actions they are taking," 63 percent believe "governments are failing young people across the world," and 56 percent believe "government is betraying them and/or future generations" (Hickman et al. 2021).

These beliefs predict a considerable amount about the behavior of youth in the climate movement. The alienation they feel signals opposition to elected and appointed authorities or at least deep skepticism toward them. The youth do not believe that the authorities are doing enough to address the climate crisis and they believe that they are the ones who will have to deal with the crisis for the rest of their lives. In some youth this alienation is presenting as utmost frustration. They are tired of asking nicely and being ignored accordingly they have adopted social action tactics which mirror their deeply held beliefs.

In 2015, a lawsuit against the United States was filed by Our Children's Trust, a coalition of youth and environmental lawyers, with the U.S. District Court of the District of Oregon. The claim argued that the U.S. Constitution guarantees the unenumerated right of a stable climate and sought the relief of the court to order the federal government to prepare and implement a plan to phase out fossil fuels to stabilize the climate. After many years of counter filings, the case was finally settled in 2020 with a ruling that stated that imposing such a remedy would exceed the authority of the courts (na 2021).

The U.S. youth movement organized under the auspices of the nonprofit Sunrise Movement in 2017 with the initial goal of electing climate crisis advocates to office in 2018. One of the main goals of the Sunrise Movement is to promote the Green New Deal to both fight climate change and to produce green jobs. They endorse candidates that support their goals and seek to get them elected (Sunrise Movement n.d.). Frustrated by President Biden's failure to pass his bold climate agenda which included many aspects of the Green New Deal, five members of the Sunrise Movement launched a hunger strike outside of the White House in October of 2021 to further pressure politicians to adopt a bold climate plan (Smith 2021).

The U.S. youth movement was bolstered by events in Europe with the protests begun by then 15-year-old Swedish protester Greta Thunberg. Beginning in 2018, Thunberg and other young people protested outside of the Swedish parliament every day for three weeks. They used social media to spread the word and news of the protest rapidly traveled around the world. In August 2018 they formed Fridays for Future, a youth-organized and -led movement calling for strikes to demand action on climate change. Non-violent civil disobedience is called for every Friday in front of local government buildings to raise awareness of the lack of sufficient action being taken on climate change (Fridays for Future 2018).

Fridays for Future quickly spread to the U.S. with the organization of Fridays for Future U.S. which developed many local chapters distributed across the country. The U.S. demands include declaring a national climate emergency and moving rapidly to green energy, investing in local communities with equity in mind, and combating systems of oppression through social justice (Fridays for Future U.S. 2022). Fridays for Future organized the First Global Climate Strike in March 2019 during which youth walked out of schools in protest of climate inaction. Another coordinated strike was launched in May 2019 (Fisher 2019). Subsequent global strikes were called by Fridays for Future (Fridays for Future 2018).

On July 12, 2019, many U.S. activists joined the Youth Climate Summit held in Miami. The summit built on the momentum of the Fridays for Future global youth movement, which focused on school strikes to demand action on climate change. In the U.S., these young climate activists were drawn heavily from the anti–Trump movement that emerged with his election in 2016 and other progressive movements that emphasize women's rights, gun control, and climate change (Fisher 2019).

U.S. youth have organized many groups to deal with climate change or other issues of importance. For instance, Isra Hirsi (16 years old in 2019) became the co-founder of U.S. Youth Climate Strike, a group organized to lead climate strikes in the U.S. Alexandria Villasenor (14 years old in 2019) is the founder of Earth Uprising, a nonprofit urging youth worldwide to rise up against climate change. Xiuhtezcatl Martinez (19 years old in 2019) organized Earth Guardians, an organization that trains youth to use civic engagement and the arts to help solve environmental issues. Jamie Margolin (17 years old in 2019) founded Zero Hour, a youth- and women of color-led movement that trains new activists to fight climate change and protect natural resources. Jerome Foster II (16 years old in 2019) founded and edited *The Climate Reporter.* He also started OneMillionOfUs, which aims to register and empower young voters for the 2020 election. Haven Coleman (13 years old in 2019) is co-founder and co-executive director of U.S. Youth Climate Strike (Schnaidt and Galvez-Shorts 2019).

The youth movement, like most other social movements, was thrown into disarray by the pandemic. American youth quickly moved away from civil disobedience protests to online spaces. With schools shut down and learning becoming virtual in 2020 and beyond, the school strike movement shut down. The Sunrise Movement turned to Zoom to begin organizing Wide Awake actions. Wide Awake actions drew from a radical youth abolitionist tactic created in the run-up to the Civil War. Wide Awake in that era confronted anti-abolitionists by night by noisily banging pots and pans outside their homes.

In a similar manner, the Sunrise Movement organized Wide Awake events targeting U.S. politicians, including senators Mitch McConnell, Lindsey Graham, and Pat Toomey (Lakhani 2020). The activists also adopted new strategies to highlight the cause, including digital protests including at least one global Fridays for Future strike held partly online in March of 2021. But as schools and society reopened, the youth once again took up the strikes. The first major Fridays for Future strike of the pandemic took place on September of 2021, although it did not rise to the size of the rallies held in 2019 when millions of young people took to the streets in the largest climate change protest in history (Adam and Noack 2021).

The Role of the Media

The media are influential actors in climate change policy and politics. It often carries stories that promote environmental action by creating awareness on the part of the public to issues that may affect their health or well-being. The media can also be a powerful force in inhibiting environmental action when, under the excuse of presenting the audience with a balanced narrative, they provide undue time and credit to views that do not represent legitimate scientific perspectives. Opponents of taking action to confront climate change, for instance, have been successful in distributing their message in this way. The wide array of media outlets includes social media, broadcast media, news magazines, and newspapers. They provide many paths for environmental issues to gain attention. Media coverage has the power to focus the public and politicians' attention on an issue which may result in some action or response, thus influencing agenda setting. The media also has a powerful role to play in framing issues and constructing narratives through which the public and politicians will view affairs (Crow and Lawlor 2016).

The Role of the Opposition

Particularly within the U.S., the role of the opposition cannot be overlooked. Opposition to environmentalism, in general, has a long history in the U.S. Environmental historian Samuel P. Hays traces the sources of the opposition and its evolution over time. Hays argues that the roots of the opposition rest with those that would defend the older economic, social, and political culture for whom newly emergent environmental values represented a threat. These interests were largely represented by traditional American farming and grazing, lumbering, and mining. The other root lay in the rise of new economic interests that considered environmental restrictions to be a threat to their activities, including manufacturing industries that processed and refined materials including steel, wood, and agricultural commodities.

Deeply tied to extraction and traditional economic production, the permanent opposition emerged to defend their "way of life" against the new environmental interests focused on improvement of quality of life. The expanding manufacturing industries of the 20th century, especially the chemical industry, joined the opposition. They were supported by land developers, who constituted another major segment of the environmental opposition. These interests formed to oppose the new waves of environmental legislation that began sweeping the country in the 1970s. Much of the role played by these groups was through political contributions to politicians they could sway to support anti-environmental ideas and administrative lobbying of government agencies with the power to implement environmental laws through the rule making process (Hays 2000).

With the election of Ronald Reagan to the White House, the last two decades of the 20th century became a rallying cry on the part of Republicans for reduction of regulations. Many of these regulations were environmental. With Reagan, the Republican party began its realignment into the party of deregulation. With this stand, the Republican party was soon identified as the anti-environmental party.

The role of opposition to climate change policy, therefore, was in no way new or a departure from many decades of environmental opposition. The Republican party added climate change as another environmental interest to oppose. Support of Republican politicians by the fossil fuel interests accompanied this process. When the oil industry (supported by the coal and gas sectors) began its disinformation campaign against climate change science, Republican politicians were eager to go along, and many adopted the political position of climate denial. George W. Bush also adopted a climate denialism stance after taking the presidency. Later in his presidency he did reverse this position but argued that doing anything about climate change would be too costly. Adaptation to living in a warmer world became his solution (Rahm, U.S. Environmental Policy: Domestic and Global Perspectives 2019).

With the election of Donald Trump, climate denialism was elevated to a new height. Trump's rhetoric included frequent claims that climate change was a hoax created by China, and during his term in office he worked insidiously to reverse U.S. environmental protections. Trump's legacy included pulling the U.S. out of the Paris Agreement, an act reversed by his successor, Joe Biden (Tharoor 2019).

Participation by International Organizations and Groups

The Role of the United Nations

The United Nations and key actors within it are central players in world environmental issues, including climate change. The UN was created in the aftermath of the Second World War as a global security institution. It is organized into a General Assembly, representing the 193 member states, and the Security Council, which consists of representative from 15 states including the permanent representatives from China, France, Russia, the United States, and the United Kingdom. From its foundation, and through the Cold War, the UN focused primarily on keeping world peace.

In the 1950s, 1960s, and early 1970s the UN began to focus on global scientific concerns that moved it closer to a direct interest in the environment. For instance, the World Meteorological Organization (WMO) got involved with the 1956–7 International Geophysical Year as well as the Global Atmospheric Research Program (O'Neill 2015). The Food and Agricultural Organization (FAO) was concerned with scientific issues surrounding food production. The World Health Organization (WHO) maintained an interest in human health and was concerned with correlation between exposure to pollution and health effects. In 1972, the UN turned to a direct interest in the environment.

UN Conferences

The world's first major meeting that focused on the environment, the United Nations Conference on the Human Environment, was held in Stockholm, Sweden, in 1972. A Declaration resulted from the Stockholm Conference elaborating many fundamental principles that would become central to the emerging environmental movement. These include the idea that human well-being depends on our environment, that humans have an obligation to protect the environment, that in developing countries most environmental problems emerge from lack of development therefore the primary obligation of developing countries is to develop, that in developed countries most environmental problems result from industrial processes and technology so developed countries have an obligation to reduce pollution, and that developed countries should devote efforts to assisting developing nations to progress in a sustainable manner.

The Declaration spoke to the need to protect and improve the environment for current populations and future generations, that humans have the responsibility to protect wildlife and habitat, that the discharge of toxic substances must be controlled, that steps must be taken to protect the seas from pollution, and that environmental education, science, and technology should be widely deployed (Rahm, U.S. Environmental Policy: Domestic and Global Perspectives 2019). Many of these key principles would be revisited in future UN conferences. The Stockholm Conference also underscored the important role that the United Nations institutions would play in global environmental governance.

Since the Stockholm Conference, the UN has hosted a series of global environmental conferences. These conferences have served as a critical mechanism to bring together world governmental and civil sector leaders to discuss policy on specific environmental issues or regimes. These conferences have resulted in elaboration of international expectations often expressed as conference declarations. These meetings have also resulted in the creation of specific UN institutions focused on the environment and development. The first of these institutions, the United Nations Environment Programme (UNEP), was created by the Stockholm Conference. Since 1972, the UN environmental conferences and agencies have been key players in the making of multilateral environmental agreements (MEAs) (Andresen 2007).

Major UN conferences on the environment and sustainable

development have been held since the UN Conference on the Human Environment in 1972. In 1987, the World Commission on Environment and Development drafted a report to the General Assembly called *Our Common Future*, also known as the Brundtland Report, which created a definition of sustainable development. The UN Conference on Environment and Development, also called the Earth Summit, was held in Rio de Janeiro in 1992. It created the Commission on Sustainable Development (CSD). The Earth Summit also created the Rio Declaration which, like the Stockholm Declaration, contained a list of principles and obligations for states to adopt and follow.

Another product of the Rio meeting was Agenda 21, a global action plan to stimulate sustainable development. Two very important MEAs were created by the Rio meeting: the UN Framework Convention on Climate Change (UNFCCC) and the Convention on Biological Diversity (CBD). Under the direction of the secretariat of the UNFCCC and various other working groups, a series of conferences have been held to address the issue of climate change since the UNFCCC took effect in 1994 (after ratification by enough parties to the agreement). The two major agreements negotiated have been the Kyoto Protocol in 1997 and the Paris Agreement in 2015.

The CBD, also signed at Rio, promotes the conservation of biological diversity and the sustainable use of Earth's biological resources. Rio's Agenda 21 was reviewed at the World Summit on Sustainable Development held in Johannesburg, South Africa in 2002. This meeting was also known as Rio+10. The UN Conference on Sustainable Development, also known as Rio+20, held again in Rio de Janeiro in 2012, produced a paper called *The Future We Want*, reformed UNEP's institutional form, and set up a process to replace CSD with a High-Level Political Forum (HLPF) for sustainable development (Rahm, U.S. Environmental Policy: Domestic and Global Perspectives 2019).

United Nations Environment Programme (UNEP)

The UNEP was supposed to be the anchor organization in the UN system for environmental issues. However, the design of the UNEP did not structure it as an organization well-equipped for centralized administration of environmental issues. Rather, the designers of UNEP saw it more as a flexible integrative body that would influence the already functioning UN organizations that had control over traditional UN policy areas including agriculture, health, labor, transportation, and industrial development. The UNEP also ended up receiving inadequate funding and, for political reasons, was located in Nairobi, Kenya. The Nairobi location worked against UNEP fulfilling its role of integration as the other organizations with overlapping policy jurisdictions were not similarly located.

The group's founders tasked it with a progression of functions to use to grapple with the world's environmental problems. Its mandate is fourfold: problem identification using scientific data, establishing policy goals and methodology, coordinating environmental action within the UN system, and building national institutional capacity. The UNEP is generally seen as unsuccessful at fulfilling this mandate fully. Rather than becoming the single leading environmental organization in the UN system, UNEP is often viewed as just one of many actors in a fragmented system (Ivanova, UNEP in Global Environmental Governance: Design, Leadership, Location 2010).

These failings were addressed to some extent in Rio+20. At that conference, UNEP's Governing Council was expanded to include all the member nations of the UN, in essence transforming it into the UN Environmental Assembly. This change is likely to give its actions more legitimacy. Rio+20 also put UNEP in a better financial position by increasing its budget. Together these changes greatly enhanced UNEP's ability to become the central coordinating body for environmental issues in the UN system (Ivanova, The Contested Legacy of Rio+20 2013).

The UN High-Level Political Forum on Sustainable Development

The Commission on Sustainable Development (CSD) was created by the Rio Earth Summit to implement the outcomes of the conference. After some early successes, the CSD became increasingly unable to accomplish its mission of turning discussions into actions. Dissatisfaction with CSD led to the establishment of the High-Level Political Forum (HLPF) after much discussion regarding other possibilities such as creating an independent sustainable development council. Efforts to develop such a council failed. At the suggestion of the G77/China, a state-led forum in which developing countries would have control was considered.

In the end, compromises resulted in the HLPF's organizational form. It is an inter-state forum that meets under the auspices of the UN General Assembly, at the head of state level, every four years. It also meets annually at the ministerial level under the auspices of the United Nations Economic and Social Council (ECOSOC). This hybrid structure and "forum" status forces HLPF to adopt an orchestration strategy—a mode of indirect or soft governance in which the orchestrator (HLPF) enlists intermediary actors (non-governmental organizations and other UN organizations) to bring a third set of actors (the world's nations) in line with its goals. The HLPF has been given ambitious goals including providing political leadership for action on sustainable development, setting the sustainable development agenda, coordinating the agenda across the UN system, and following up on the progress in implementing UN sustainable development goals. It does this with modest resources. While HLPF has legitimacy and political prestige which puts it in a good position to bring leadership within the UN system, it also must compete with many other UN system organizations that deal with sustainable development (Abbott and Berstein 2015). The HLPF is the key organization within the UN that manages the 2030 Agenda for Sustainable Development which the UN sees and inseparable from the climate agenda (United Nations Sustainable Development and Climate Agenda 2022).

The Global Environment Facility

Other international environmental organizations exist. For instance, the Global Environmental Facility (GEF), established just before the Earth Summit, is a unique organization that works with many partners to deal with pressing global environmental issues. Partners of the GEF include developed and developing nations, non-governmental organizations (NGOs), and agencies of the United Nations. The GEF is administered jointly by the UNEP and the United Nations Development Programme (UNDP) with funding coordinated by the World Bank. One important role of GEF is to make funds available to developing nations and to provide capacity building so that they can fulfill their requirements under several international environmental agreements, such as: the Minamata Convention on Mercury, the Stockholm Convention on Persistent Organic Pollutants, the United Nations Convention on Biological Diversity, the United Nations Convention to Combat Desertification, and the United Nations Framework Convention on Climate Change (O'Neill 2015).

Environmental Regime Secretariats and Environmental Departments in Non-Environmental Organizations

A good way to envision the organization of environmental programs in the United Nations is to consider each major area of environmental concern, for which an international treaty has been negotiated, a regime. For instance, there are regimes for ozone, hazardous waste, toxic chemicals, ocean pollution, climate, transboundary air pollution, global biodiversity, endangered species, wetlands, desertification, and whaling. Each environmental regime has its own secretariat, which is a permanent international governmental body with a permanent staff that reports to the regime's Conference of Parties. Many of the secretariats are located within UNEP, but some, like the secretariat for the UN Framework Convention on Climate Change (UNFCCC), are managed directly by the UN. Others, such as the toxic waste regime secretariat exist outside of the UN completely (for example, the Ramsar Convention secretariat). Many treaty-based environmental regimes have scientific and technical support organizations. For instance, the UNFCCC has the Intergovernmental Panel on Climate Change (IPCC), which advises the climate change regime on new scientific findings periodically (O'Neill 2015).

In addition to the secretariats of international environmental treaties, there are other international governmental institutions to consider. Specifically, these include environmental departments located within international organizations that do more than just environmental policy. These would include such units as environmental departments and subdivisions of the World Bank, the environmental department of the secretariat of the International Maritime Organization, and the environmental directorate of the Organization for Economic Cooperation and Development (OECD) secretariat (Biermann and Siebenhener 2009). It is also important to remember that many groups with a wider mission, such as the International Monetary Fund (IMF) or the World Trade Organization (WTO), have enormous influence on environmental issues whether or not they have set up a dedicated bureaucracy within their organizations to handle specific environmental issues.

International Environmental Organizations

Environmental non-governmental organizations (NGOs) are plentiful in the international arena. Some of the more prominent ones are discussed here. They differ in two primary ways. First, many are older organizations and thus had a standing mission when climate change became an issue. As a result, they added some aspect of climate advocacy to their standing mission rather than adopting climate change as their key mission. Second, they vary regarding their tactics. Some are more adversarial or confrontational in their approaches while others are more collaborative and cooperative.

Greenpeace International oversees a network of regional and national Greenpeace organizations that work with local communities to address a host of environmental issues. Greenpeace International has chapters in 26 regions or countries. Greenpeace began in 1971 with the mission of stopping above ground nuclear testing. Over the years it has adopted many other missions including protection of the oceans, forests, and whales; exposure of toxic dumping; protection of the ozone layer; opposition to genetically modified foods; and climate change activism (Greenpeace International 2022).

Greenpeace differs from other environmental NGOs primarily in the tactics used. Greenpeace uses direct action tactics—or, as it phrases it, "non-violent creative action"—to achieve its goals. These actions include sailing ships between whalers and whales to stop a hunt, hanging large posters from public buildings, and blocking mega ships from European ports to protest the destruction of the rain forest undertaken to grow the crop they carry. Such actions are meant to be provocative and to draw attention of the press, public, and government authorities. Greenpeace activists hold loud public protests to advocate for their goals. In terms of climate, Greenpeace advocates what they call "Real Zero" as opposed to net zero, which they argue allows polluters to use carbon offsets to continue polluting (Greenpeace 2022).

Fauna and Flora International (FFI) works to conserve threatened species and ecosystems worldwide. It is the one of the oldest international environmental NGOs, tracing its foundation to 1903. It works in more than 40 countries in Africa, Asia-Pacific, and the Americas. While FFI focuses on biodiversity, it is keenly aware that biodiversity is threatened by climate change; it believes that the complex nature of biodiversity requires a multisector approach to solutions, therefore, FFI's mainstream tactic is to partner with businesses, governments, and NGOs (Fauna and Flora International 2022).

Another older international environmental organization is Nature Friends International (NFI), which was formed in 1895 and has today more than 350,000 members who work in active local groups of sections. An umbrella organization of 45 national member organizations, NFI is amongst the ten largest European environmental NGOs. They are committed to ecological and sociopolitical causes. They advocate for free access to nature for all people, for environmental and socially just tourism and leisure activities, sustainable tourism, nature education, diversity and intercultural exchanges. Their climate activities are organized around the concept of climate justice which supports the right of every person to experience a stable climate. NFI supports the idea that developed nations contribute more to the problem than developing nations and so advocates for transfer of money are resources from the developed world to the developing world in order to create climate justice (Nature Friends International 2022).

The International Union for Conservation of Nature (IUCN) was established in 1948 in France. It works to facilitate partnerships between governments and NGOs with the shared goals of protecting nature, promoting international cooperation, and providing scientific knowledge and tools to guide conservation actions. The IUCN's early advocacy focused on raising awareness of the damaging effects of pesticides on biodiversity and protecting habitat. Beginning in the 1970s, IUCN partnered with the United Nations to work on the Ramsar Convention on Wetlands, the World Heritage Convention, the Convention on International Trade in Endangered Species, the Convention on Biological Diversity, the United Nations Framework Convention on Climate Change, and the United Nations Convention on Desertification. In the early 2000s, IUCN developed a business engagement strategy. By 2022 it had more than 1,300 members including states, government agencies, and NGOs, plus more than 15,000 international experts. It advocates for nature-based solutions as key to the implementation of international agreements, such as the Paris climate change agreement (International Union for the Conservation of Nature 2022).

The World Wildlife Fund (WWF) was established in 1961 by a group of individuals that sought to secure funding to undertake projects to protect habitat and species threatened by human development. The group was motivated in part by the financial difficulties facing IUCN and felt that a new fundraising initiative might help IUCN and other conservation groups carry out their missions. A member of IUCN's executive board later became WWF's first vice president. The group established three national organizations in the UK, Switzerland, and the U.S. It works in more than 100 countries today, partnering with governments, businesses, and NGOs to work for conservation of nature. The WWF's work is focused on six topics each with its own goals. These include: to create a climate-resilient world powered by green energy, to double net food availability and freeze its footprint, to conserve the world's forests, to secure freshwater for people and nature, to safeguard the oceans and marine life, and to preserve wildlife and habitat (World Wildlife Fund 2022).

The Natural Resources Defense Council (NRDC) works with scientist and advocates for a clean and stable environment to achieve its goals. Founded in 1970, NRDC is a U.S.-based international environmental advocacy organization with more than three million members. It partners with governments, businesses, and NGOs as its primary strategy to implement programs. To address the climate crisis, it specifically promotes clean energy, energy efficiency, and electric vehicles (NRDC 2022a).

While many of the international NGOs came into being before climate change and simply added a climate change mission to their list of activities, several international NGOs formed specifically to deal with the climate crisis. One is 350.org, which was founded in 2008 by a group of U.S. university students and writer Bill McKibben, the author of *The End of Nature*. The name refers to the safe level of parts per million of carbon dioxide in the atmosphere. 350.org organized the International Day of Climate Action in 2009 and other annual action activities in succeeding years. Like most climate activist movements, 350.org fights not only for a livable planet but also for one with equity (350.org 2009). Another is Extinction Rebellion (XR), initially formed in the UK in the fall of 2018, when XR brought thousands out on London streets to draw attention to the climate crisis. By December the movement had spread to 35 countries, and it has created an offshoot group called ER Youth (Yoder 2021).

A spinoff of ER is Scientist Rebellion (SR), a network of scientists of all kinds that focus on civil disobedience to bring attention to the climate crisis. Frustrated with the lack of political progress that has been made in the years since climate science has clearly indicated the human role in planetary warming, these scientists have abandoned their allegiance to neutrality and instead have adopted the use of direct action civil disobedience tactics to call attention to the need to make political progress. In April 2022, this group organized over 1,000 scientists in 26 countries to protest. Influential in the SR is Peter Kalmus, a NASA research scientist. Kalmus and three other scientists were arrested in Los Angeles during the April event for chaining themselves to the doors of JPMorgan Chase, the largest funder of fossil fuels. While chained to the doors, Kalmus gave a speech that quickly went viral on social media. In that year, Kalmus became the most followed climate scientist on Twitter with 252,000 followers and surpassed the fame of climate scientists Michael Mann and Katherine Hayhoe (Quackenbush 2022).

The Rise of the International Youth Movement

Youth worldwide are feeling great anxiety about climate change. This has been documented by a study in 10 countries (Australia, Brazil, Finland, France, India, Nigeria, the Philippines, Portugal, the UK, and the U.S.) of 10,000 people aged 16–25. The study surveyed the youth between May 18 and June 7, 2021. Data were collected on the participants' thoughts about climate change and attitudes toward governments' actions on climate change. The results showed that respondents in all countries were worried about climate change, with 59 percent saying they were worried or very worried and at least 84 percent were moderately worried.

More than half of the youth in the study reported feelings of sadness, anxiousness, anger, powerlessness, helplessness, and guilt. More than 45 percent said their feelings about climate change negatively affect the functioning of daily life. Three-quarters of the youth reported that they think the future is frightening and 83 percent said they think people have failed to take care of the planet. The respondents rated governmental response to climate change negatively and reported a greater feeling of betrayal than reassurance. The perceived failure of governments to respond to the climate crisis is associated with increased distress among youth (Hickman et al. 2021). These emotions have given rise to the international youth movement.

The youth movement's first global strike took place March 15, 2019, in more than 100 countries with more than 1,600 separate events following in the wake of weekly strikes, mainly in Western Europe, the United States and Australia but also in poorer nations including India, Colombia, and Uganda. The global action was unprecedented insofar as it was mobilized by schoolchildren. In an open letter, the global coordinating group demanded that all countries need to meet their commitments stated in the Paris agreement. They also called for climate justice for all the future victims of climate change (Adman and Uba 2019).

The significance of schoolchildren leading the movement is noteworthy as that role in the past has often fallen to college students. It was college students that played a pivotal role in the U.S. civil rights movement, student protests with French police in 1968, and in 1989 in Tiananmen Square demonstrations in China. Older students have often been the driving force of protest movements. But this youth-led climate movement emerged because of the leadership of Greta Thunberg. At age 15, Thunberg began a climate strike in front of the Swedish parliament. The outcome of this climate strike was the creation of Fridays for Future. The youth that she inspired began to organize themselves in a grass-roots, bottom-up fashion focusing on building support in their school classes and by using social networking. Youth overwhelmingly see climate change as a threat to their futures and Greta Thunberg became a symbol of the movement. She immediately was invited by adult advocates of climate action to meet with them (Adman and Uba 2019).

In September of 2019 she and a group of youth activists were invited to meet with the members of the U.S. Senate's climate crisis task force, led by Senator Ed Markey, co-sponsor of the Green New Deal plan. Thunberg arrived in the U.S. after crossing the Atlantic on a solar-powered yacht (Gambino 2019). She arrived in the U.S. to speak at the United Nations Climate Action Summit on September 23, 2019, along with Alexandria Villasenor, a U.S. Fridays for Future striker and founder of Earth Uprising (Nargi 2019). The first global climate strike was designed to be held at the same time as the meeting of the United Nations. That strike drew millions of participants worldwide and is thought to be the largest climate change protest in history to date (Adam and Noack 2021).

The global climate strikes continued into the pandemic but were disrupted by the shutdowns that affected most countries. They resumed in 2021 in the lead-up to COP26, held in Glasgow, Scotland, beginning October 31, 2021. The Glasgow talks pointed to the divide between the international authorities and the climate activists. Of the 130 heads of state that attended the Glasgow talks, fewer than 10 were women and their median age was 60. In contrast, the protests that poured into the streets of Glasgow were led by mainly female youth climate activists, many too young to vote in their home countries. The organizers of the Glasgow meeting included youth activists as speakers and many of the government authorities present went to great lengths to assure attendees that they had heard the voice of the youth. Yet the commitments coming out of Glasgow were insufficient to satisfy the youth protesters.

Greta Thunberg spoke before the cheering crowds in the streets of Glasgow and declared the meeting a failure (Sengupta, Young Women Are Leading Climate Protests. Guess Who Runs Global Talks? 2021). Thunberg's comments at Glasgow echoed her earlier comments at Youth4Climate summit held in Milan, Italy late in September 2021. There she dismissed the empty promises of global leaders as "blah, blah, blah" (Carrington 2021).

The Opposition Worldwide

Climate change denialism and efforts to stop policy action aimed at reducing global warming is not just a U.S. phenomenon, although it is mostly associated with the U.S. Globally there have been several national leaders who have played a role in exacerbating the climate crisis though their efforts to deny the worst effects of climate change and to promote national policies whose main effects significantly worsen global warming. Key among these was Jair Bolsonaro, Brazil's president from 2019 to 2022. Bolsonaro's policy toward the Amazon rainforest, the world's largest rainforest, will have global consequences. Bolsonaro's policies fostered the continued deforestation of the Amazon, which has suffered from loss of forest for the last three decades of the 20th century and the first two of the 21st century. Scientific estimates suggest that nearly 17 percent of the Amazon has been lost during this time period and that in the 2020s such trends were accelerating.

Much of this quickening is due to Bolsonaro's policies that promoted cattle ranching and soybean production. Both activities created incentives for farmers and ranchers to burn large swaths of the rainforest to clear land. By 2020, Brazil was one of the world's top exporters of beef and soy. Bolsonaro's policies to dismantle protections for indigenous communities, who have lived in harmony with the land for centuries, further deepened the problem. By promoting policies that aid in the commercialization of the Amazon, Bolsonaro forced many indigenous communities to move, subjecting his administration to indigenous-led protest movements across Brazil.

Bolsonaro's policies also resulted in widespread fires breaking out across the Amazon in 2019. The Group of Seven (G7) sought to provide aid to stop the fires but Bolsonaro refused it and accused the G7 of infringing on Brazilian sovereignty (Roy 2022). Bolsonaro, in his first two years in office, presided over destruction of 10,000 square miles of the Amazon. His stance on climate denial earned him the nicknames "Trump of the Tropics" and "Captain Chainsaw." The importance of the Amazon as a carbon sink made Bolsonaro's policies critical to reaching the goals of the Paris Agreement. Scientists argue that without ending deforestation, particularly of the "lungs of the Earth," the fight to limit warming to 1.5 degrees Celsius will be lost (Goodell 2021).

During COP26 held in Glasgow, Scotland, in 2021, discussions on achieving the Paris Agreement's goals were disrupted by a newly emerging form of climate denialism. While not openly challenging the scientific consensus on climate change, delegates from several countries did demonstrate a new tactic—that of nudging or redirecting climate policy to minimize its impacts on the fossil fuel sector. Australia, Saudi Arabia, and Japan each requested of the UN that it downplay the need to move rapidly away from fossil fuels. India and China later joined in the pressure. The fossil fuel sector also played an outsized role with more than 503 fossil fuel lobbyists accredited to be part of the meeting, more than the number of delegates allowed to any single country. The text that came out of the COP26 meeting did directly mention fossil fuels, but the wording was weak, largely as a result of the influence of the fossil fuel sector (Tindall, Stoddart and Dunlap 2021).

Conclusion

The climate change debate includes both domestic U.S. and global actors. For the U.S., the debate is dominated by the federal government, state and local governments, and the U.S. court system. The U.S. nonprofit sector plays an important role in the policy debate as well. Many of these nonprofits are U.S.-based only, while others are U.S.-based chapters of international non-governmental organizations. Within the U.S. and internationally, the climate change debate since the mid–2010s has been dominated by youth activists who have had an outsized role on both public opinion, largely through media coverage, and attempts to sway policy action.

Climate change is a politically charged issue in the U.S. in particular and efforts to reduce emissions has drawn opposition from many politicians, climate deniers, and industries fearing loss of profits if their activities are restricted. Internationally, efforts to confront climate change largely rest with the UN System; its many conferences; the climate secretariat; and various other organizations formed or supported by the UN including the UNEP, the UN High-Level Political Forum on Sustainable Development, the GEF, the IPCC, international environmental organizations, and the global youth movement. Global opposition to policy actions taken to slow or reverse global warming have been ongoing in all of the 21st century so far and have been dominated by the fossil fuel sector and those nations or groups expecting to achieve some benefit from the sector.

CHAPTER 3

Inherited 20th-Century Climate Change Policy

Introduction

The 21st century inherited a great deal of climate change policy from the 20th century. This chapter discusses the key developments that arose before the beginning of the 21st century, which corresponded with the inauguration of George W. Bush as president in 2001. Most of the 20th-century policies of significance happened during the terms of George H.W. Bush, who served from 1989 to 1993, and Bill Clinton, who served from 1993 to 2001.

The science legacy has a longer history—in fact, early scientific understanding of global warming began in the early 1800s. It would not become clear to scientists until much later that the burning of fossil fuels associated with the explosion of new technologies during the industrial revolution would pose a threat to the planet. But by the turn of the 21st century, the vast scientific understanding of just how much a threat climate change posed would become understood.

Early understanding of this threat created a global political response to climate change driven by the United Nations and some of its associated groups. The world's first treaty on climate change came into force in the 20th century. However, the stirrings of political discord, at least within the United States, were also rising. By the time President Clinton negotiated the second climate treaty, the Kyoto Protocol, the political divisions within the U.S. were clear. Opposition to climate action grew.

This chapter discusses the international and U.S. policies that emerged following the clarification of the science behind global warming. Discussed first is the science inheritance and then the policies that emerged. It closes with a discussion of the opposition, including the fossil fuel industry, conservative deregulatory politicians, and Christian conservatives associated with the Moral Majority and the Christian coalition.

The Science Legacy

Early efforts to understand how the Earth is a habitable planet go back to the 19th century. Joseph Fourier, in 1824, suggested the greenhouse effect. He calculated that a planet the size of Earth, with Earth's distance from the Sun, should be much colder. He concluded therefore that something was acting as an insulating blanket to warm the planet. In 1856, carbon dioxide and water vapor were identified by Eunice Foote as the agents of warming. She showed that together they trapped infrared (heat) radiation from escaping back into space, thus making the planet warmer. In the 1860s, physicist John Tyndall explained the Earth's natural greenhouse effect and showed that changes in the atmosphere could bring about climate variations. In 1896, Swedish scientist Svante Arrhenius predicted that changes in atmospheric carbon dioxide levels could alter surface temperature through the greenhouse effect (NASA 2022). Each of these scientific advances laid the early foundations for understanding how our planet was kept at a livable temperature.

Scientific efforts to further explain the greenhouse effect continued into the 20th century. In 1939, Guy Callendar tied carbon dioxide increases in the atmosphere to global warming. In 1956, Gilbert Plass formulated the Carbon Dioxide Theory of Climate Change (NASA 2022). Plass, a Johns Hopkins physicist, told *Time* magazine that the rate that industry was pumping carbon dioxide into the atmosphere would mean that temperatures on the planet would rise 1.5 degrees Fahrenheit every 100 years (Time 1953).

Beginning in 1958, Charles Keeling began recording carbon dioxide levels in the atmosphere in Mauna Loa, Hawaii. Keeling's time series data provided evidence allowing scientists to differentiate between carbon dioxide coming from fossil fuel use and carbon dioxide coming from the natural carbon cycle. Keeling's data would later be instrumental in showing the impacts of fossil fuel use on global warming. By the 1960s, scientists verified that atmospheric carbon dioxide levels were increasing annually, and they were able to predict that atmospheric temperatures might rise by several degrees by 2060. Such predictions were troubling as scientists began to ask questions about how quickly plants and animals could adapt to such warming. These discoveries also corresponded to a larger social and political movement emerging in the U.S. regarding the general environment and how human actions affected it (Rahm 2010).

Early Awareness of Global Warming

A great deal of uncertainty entered the debate over global warming in the 1970s because of the impact of small particles or aerosols on the atmosphere. These particles had the effect of blocking sunlight and reducing temperatures. Scientists in the 1970s even suggested that the planet was set for another Ice Age. Historical events of the decade, though, did focus attention on fossil fuels. The Organization of Petroleum Exporting Countries (OPEC) boycotts of the 1970s led to increased use of coal, the most polluting of all the fossil fuels, and this led to concerns of what the excess release of carbon dioxide would do. President Carter called on Congress to allocate funds to study the problem.

The Departments of Defense (DOD) and Agriculture (USDA) along with the National Oceanic and Atmospheric Administration (NOAA) reached out for scientific opinions on climate change. They surveyed 24 climatologists regarding whether they thought the Earth was undergoing a warming or a cooling trend, but results were mixed. This lack of consensus led to a call for more research. A new database of climate indicators was developed and data from ships and satellites were collected. These new data revealed that climate was a complex system with many feedback loops and that climate responded to many inputs. These included volcanic eruptions, solar flares, ocean currents, cloud cover, and the role of atmospheric gases besides carbon dioxide.

This new evidence created a call from many scientists in the late 1970s and early 1980s that the world should take action to reduce greenhouse gas (GHG) emissions. In 1981, a report from the National Aeronautics and Space Administration (NASA) predicted global warming trends that might melt the West Antarctic Ice Sheet, thus raising sea-levels considerably. This was followed by a 1983 report from the Environmental Protection Agency (EPA) which predicted that global temperatures could rise 2 degrees Celsius by 2050. That same year, the National Academy of Sciences released a report agreeing with the EPA report (Rahm 2010).

The summer of 1988 was the hottest year then on record and it was made politically controversial by the appearance of Dr. James Hansen of NASA at a Congressional hearing. Hansen told the Congressional committee that the evidence was clear that climate change had begun, and that action should be taken to reduce GHG emissions. Prior to Hansen's testimony, politicians and most individuals who thought about global warming saw it as a future event. This allowed them to hesitate while considering taking actions that might have present economic ramifications. With Hansen's testimony, a new urgency emerged in the U.S. climate debate that was matched by international events (Rich 2019).

The Growth of Global Concern: The Establishment of the Intergovernmental Panel on Climate Change

Concern about global warming was a worldwide phenomenon in the same year Dr. Hansen testified before Congress. The Intergovernmental Panel on Climate Change (IPCC) was created to assess the science on climate change. It was created in 1988 by the World Meteorological Organization (WMO) and the United Nations Environment Programme (UNEP) to provide scientific advice to the world's governments on policies they might enact to address climate change. The IPCC provides regular assessments of climate change impacts, its future risks, and options for mitigation and adaptation. Thousands of scientists worldwide volunteer their time to review the thousands of papers published each year on climate change and to create the IPCC assessments. The IPCC provides analysis on the strengths of the scientific knowledge and indicates where further research is needed; it does not, itself, do scientific research. The IPCC is divided into three working groups: the physical science basis of climate change, climate change impacts, and mitigation of climate change. The IPCC also has a task force on greenhouse gas inventories. The main role of the task force is to devise methodologies for calculating and reporting national greenhouse gas emissions and reductions (IPCC, About the IPCC 2022).

The establishment of the IPCC by WMO and UNEP was endorsed by the United Nations General Assembly in December of 1988. The primary task was to create a report that contained a comprehensive review and recommendations with respect to the level of scientific knowledge about climate change, the social and economic impacts of climate change, and potential strategies for inclusion in a future international convention on climate. The IPCC delivered its First Assessment Report (FAR) in 1990. The report emphasized the global challenge that climate change posed and the need for international cooperation in confronting this challenge.

Between 1990 and 2001 the IPCC put out many reports and assessments that greatly increased the world's understanding of climate change and its impacts. For instance, in 1992, IPCC released another supplemental assessment. In 1994, a special report focusing on radiative forcing of climate change and an assessment of the 1992 emissions scenarios was released. Also in that year, the IPCC published an improved methodology for measuring and reporting national GHG inventories.

In 1995, the Second Assessment Report (SAR) was released, including separate reporting of the science of climate change, the impacts of climate change, and the social and economic aspects of climate change.

By 1995 knowledge had increased and with it the certainty of the reports that climate change was the result of human activity. In 1996, the IPCC again released an updated methodology on measuring national inventories. In 1997, they published a special report on regional impacts and an assessment of vulnerabilities. In 1999, IPCC released a report on aviation and its impacts. In 2000, they revisited emission scenarios and released a report on land use, land use change, and forestry. In 2001, the IPCC released its Third Assessment Report (TAR) with separate reports on the scientific basics, impacts and vulnerabilities, adaptation, and mitigation (IPCC, History of the IPCC 2022).

The more than a decade between the FAR (first report) and the TAR (1990 to 2001) had vastly increased the scientific understanding of climate change as well as its impacts on and the vulnerabilities of human settlements. With each new release, the certainty of the scientific consensus grew and the urgency to take policy action was underscored.

Early International and U.S. National Policy Action

The United Nations

The United Nations had begun to take interest in environmental issues in the 1970s with the holding of the first international conference on the environment. The United Nations Conference on the Human Environment was organized and held in Stockholm, Sweden, in 1972. Participants adopted several declarations including the Stockholm Declaration which marked the beginning of the dialog between developed and developing nations on the state of the environment and the appropriate role of each in achieving desired outcomes. One of the major accomplishments of the Stockholm meeting was the creation of UNEP, which was one of the organizations that created the IPCC (UN, Conferences/Environment and Sustainable Development/Stockholm 2022).

The next world conference on the environment was held twenty years later in Rio de Janeiro, Brazil. The 1992 United Nations Conference on Environment and Development, or the Earth Summit, had the broad goal of producing a new blueprint for international action on environment and development. A major achievement of the conference was Agenda 21, a document outlining how to achieve sustainable development into the 21st century. Another major achievement was the Rio Declaration with its 27 principles. The Earth Summit also produced the

United Nations Framework Convention on Climate Change (UNFCC), the Convention on Biological Diversity, and the Convention of Desertification (UN, Conferences/Environment and Sustainable Development/ Rio 2022).

Several of the principles of the Rio Declaration are important to understand, as they provided the future framework for international relations on the environment and sustainable development. The first of these was Principle 2 which said,

> States have, in accordance with the Charter of the United Nations and the principles of international law, the sovereign right to exploit their own resources pursuant to their own environmental and developmental policies, and the responsibility to ensure that activities within their jurisdiction or control do not cause damage to the environment of other States or of areas beyond the limits of national jurisdiction.

Thus, Principle 2 established the national sovereignty of countries that are home to some of the world's key ecological assets including the rainforests which act as the lungs of the planet. Unfortunately, in subsequent years political officials from several of these countries would emphasize only the aspect of national sovereignty in order to develop these regions and not the responsibility to assure that domestic actions would not spill over creating negative global outcomes.

Principle 7 outlined the differences between developed and developing nations and their common but differentiated responsibilities stating,

> States shall cooperate in a spirit of global partnership to conserve, protect and restore the health and integrity of the Earth's ecosystem. In view of the different contributions to global environmental degradation, States have common but differentiated responsibilities. The developed countries acknowledge the responsibility that they bear in the international pursuit of sustainable development in view of the pressures their societies place on the global environment and of the technologies and financial resources they command.

Thus, Principle 7 established a dichotomy of nations with the rich developed countries having to take first actions to reduce climate change pollution while developing nations could instead consider first their needs to develop.

This dichotomy was based on the concept that developed countries became rich by polluting while developing countries often caused negative environmental outcomes because of their failure to develop. For instance, a subsistence farmer on the outskirts of the rainforest needing more land for crop production would resort to slash and burn actions to gain only marginal land for additional food production. In subsequent years, poverty would drive a repetition of such environmentally unsound actions. Therefore, the framework emphasized that the developed nations had the primary responsibility to end pollution and that the developing nations had the primary responsibility to develop.

Principle 15 outlined what came to be known as the precautionary principle,

> In order to protect the environment, the precautionary approach shall be widely applied by States according to their capabilities. Where there are threats of serious or irreversible damage, lack of full scientific certainty shall not be used as a reason for postponing cost-effective measures to prevent environmental degradation [UN, Report of the United Nations Conference on Environment and Development 1993].

The precautionary approach was adopted by most nations; however, the U.S. failed to widely adopt this approach and would use the lack of certainty in the science of climate change again and again to justify inaction. These three principles became very influential and controversial in later international environmental negotiations on climate change.

The United Nations Framework Convention on Climate Change

For the issue of climate change, the most important outcome of the Rio meeting was the creation of the United Nations Framework Convention on Climate Change (UNFCCC), with the objective of

> stabilization of greenhouse gas concentrations in the atmosphere at a level that would prevent dangerous anthropogenic interference with the climate system. Such a level should be achieved within a time-frame sufficient to allow ecosystems to adapt naturally to climate change, to ensure that food production is not threatened and to enable economic development to proceed in a sustainable manner [UN, United Nations Framework Convention on Climate Change 1992].

The UNFCCC was drafted at Rio and adopted in 1994 after it was ratified by nations wishing to become parties to the agreement. For the U.S., that meant that the George H.W. Bush administration was primarily engaged in obtaining ratification.

The U.S. was a strong supporter of the treaty. President Bush personally attended the Rio Conference and signed the convention there. He submitted the convention to the Senate for ratification in 1992 and after ratification signed the instrument that same year. Doing so made the U.S. the first industrial nation in the world to adopt the treaty. In his signing statement, Bush emphasized the leading role the U.S. had played in crafting the treaty which required all parties to inventory all sources and sinks of greenhouse gases and to create a national climate change program. Calling the treaty a first step in international efforts to address climate change, Bush said the U.S. looked forward to the next conference of parties (COP) which would be scheduled after the required number of countries ratified the treaty (Bush 1992).

For the U.S., the Environmental Protection Agency began submitting the annual inventory of greenhouse gas emissions and sinks beginning in the early 1990s. This report provides a comprehensive accounting of all man-made GHGs including carbon dioxide, methane, nitrous oxide, hydrofluorocarbons, perfluorocarbons, sulfur hexafluoride, and nitrogen trifluoride. In addition, the report tracks all sinks including the uptake of carbon in forests, vegetation, and soils through land management in their current use or as land is converted to a different use (EPA 2022). In terms of the mandate in the UNFCCC to create a national climate change program, President George H.W. Bush had by presidential initiative in 1989 created the U.S. Global Change Research Program. This program was later Congressionally mandated by the Global Change Research Act of 1990, which Bush signed into law. This law established the U.S. National Climate Assessment (NCA), a report describing the impacts of climate change on the U.S. (Globalchange.gov 2022).

The UNFCCC came into force in 1994, with 196 parties. The parties met annually to assess progress and take additional steps to combat climate change. The first COP was held in Berlin in 1995, and presided over by Germany's then environmental minister, Angela Merkel (Merkel would later serve as the German Chancellor from 2005 to 2021). President Bill Clinton and Vice President Al Gore were in office in the U.S. when COP1 occurred. The parties agreed in Berlin that the UNFCCC was not strong enough to reach the goals of the convention and so in the Berlin Mandate they established a process to negotiate a strengthened treaty for developed countries, thus paving the way toward the Kyoto Protocol. The Berlin Mandate emphasized that advanced industrial countries had an obligation to take the first steps in combating climate change and underscored the importance of the separate but differentiated responsibility concept elaborated in Principle 7 of the Rio Declaration. The Kyoto Protocol would be officially adopted December 11, 1997, and it largely followed the guidelines outlined in the Berlin Mandate (UNFCCC 2020b).

The Clinton-Gore team accepted the Berlin Mandate, angering some members of Congress. As early as 1995, the House Energy and Commerce

Committee had held a hearing that specifically suggested that the U.S. not sign a treaty that did not include all major emitters. In the Senate, Clinton's acceptance of the Berlin Mandate created considerable frustration. Senator Inhofe issued a statement arguing the sentiments of the members of Congress were not being considered (Hovi, Sprinz and Bang 2010). It is not entirely clear why the Clinton-Gore team accepted the Berlin Mandate and the Kyoto Protocol knowing full well that opposition in the Senate would cause its ratification to be uncertain at best.

The Kyoto Protocol

The Kyoto Protocol had been adopted shorty after release of the IPCC's Second Assessment Report (SAR), which had come out in 1995. The SAR provided improved science and overall general information that helped with the adoption of the Kyoto Protocol (IPCC, History of the IPCC 2022). The Kyoto Protocol was adopted by the third conference of parties, on December 11, 1997, becoming the world's first greenhouse gas reduction treaty (UNFCCC 2020b).

The Kyoto Protocol drew on the logic established by Principle 7 of the Rio Declaration which recognized the common but differentiated responsibility of countries. Accordingly, the Kyoto Protocol required reductions of the six main greenhouse gases—Carbon dioxide (CO^2), methane (CH^4), nitrous oxide (N^2O), hydrofluorocarbons (HFCs), perfluorocarbons (PFCs), and sulphur hexafluoride (SF^6) for what were called the Annex I countries (which were listed in the Kyoto Protocol's Annex B). These countries included: the EU-15 (Austria, Belgium, Denmark, Finland, France, Germany, Greece, Ireland, Italy, Luxembourg, Netherlands, Portugal, Spain, Sweden, and the United Kingdom), Bulgaria, Czech Republic, Estonia, Latvia, Liechtenstein, Lithuania, Monaco, Romania, Slovakia, Slovenia, and Switzerland, the U.S., Japan, Canada, Hungary, Poland, Croatia, New Zealand, the Russian Federation, Ukraine, Norway, Australia, and Iceland.

Each party had set reduction targets for the first commitment period which ran from 2008 to 2012. These targets were measured from a baseline of 1990 emissions. They included an 8 percent reduction for EU-15 nations (but the EU-15 nations were allowed to set individual country goals that would total 8 percent reduction for the region) and Bulgaria, Czech Republic, Estonia, Latvia, Liechtenstein, Lithuania, Monaco, Romania, Slovakia, Slovenia, and Switzerland; a 7 percent reduction for the U.S.; a 6 percent reduction for Canada, Hungary, Japan, and Poland; a 5 percent reduction for Croatia; and no reductions for New Zealand, the Russian Federation, or Ukraine.

Under the terms of the agreement, Norway was allowed a 1 percent increase, Australia an 8 percent increase, and Iceland a 10 percent increase. The Kyoto Protocol required no reductions for developing nations and explicitly recognized the special needs and concerns of developing nations. It required Annex I nations to specifically report how they would meet their emission targets while not impacting developing nations. The Kyoto Protocol also established an Adaptation Fund to finance adaptation and development projects in developing countries (UNFCCC 2020a).

Clinton, the U.S. Senate, and the Kyoto Protocol

The Kyoto Protocol's reliance on the concept of separate but differentiated responsibilities would prove problematic for ratification in the U.S. While the U.S. delegation, led by known environmentalist and Vice President Al Gore, was successful in crafting many aspects of the document that finally emerged, there was no attempt to override the approach towards developing nations. Gore had a long history of promoting climate action. As a congressman, senator and vice president, Gore organized hearings on climate change and pushed for global actions. He was the U.S. delegate to the Rio Earth Summit in 1992 and played a role in the establishment of the UNFCCC. Gore also helped broker the Kyoto Protocol (Al Gore 2022).

It is widely believed that the U.S. had substantial impact on the development of the flexibility mechanisms that became part of the Kyoto Protocol. These included carbon trading, the Joint Implementation program (JI), and the Clean Development Mechanism (CDM). Carbon trading was introduced in an effort to mimic the success of the U.S. Clean Air Act Amendment's acid rain program, which introduced the concept of cap-and-trade into the mechanisms used in the U.S. regulatory system. In such a system, the pollutant is given a market value which emitters can trade to create efficiencies in their reductions. The JI program and the CDM also created flexibilities in the Kyoto Protocol. JI allowed developed nations to put in place programs in other developed nations and get credits for those programs toward their required reductions. The CDM did the same for programs created by developed nations in developing nations (Rahm 2010).

While the Kyoto negotiations were on-going, however, members of the Senate continued to express their concern that their opinions were not being heard. The Senate sent a powerful message to President Clinton with the passage of the Byrd-Hagel resolution which said that "the

United States should not be a signatory to any protocol ... which would (A) mandate new commitments to limit or reduce greenhouse gas emissions for the Annex I Parties, unless the protocol ... also mandates new specific scheduled commitments ... for Developing Country Parties within the same compliance period, or (B) result in serious harm to the economy of the United States." The Byrd-Hagel resolution was not binding—rather, it was a sense-of-the-Senate resolution—but it was passed on a 95–0 vote.

So, the question of why Clinton-Gore would negotiate and sign a treaty that they knew could not pass the Senate is an important one. The idea that Clinton-Gore were more interested in negotiating vigorously for a treaty that made them look good rather than one that the Senate would actually sign has some merit. Before Gore arrived at Kyoto, the U.S. negotiating team was fighting hard for terms for the U.S. that would include a zero percent increase in emissions over the compliance period. Such a provision would have been more appealing to the Senate than the 7 percent reduction that Clinton finally signed. It has been suggested that the Clinton-Gore team was more interested in the legacy of an environmentally tough treaty than one that would be more sellable to the Senate (Hovi, Sprinz and Bang 2010).

In the end, while Clinton signed the treaty, he never presented it to the Senate for ratification. This failure of policy not being initiated at the federal level left it to subnational governments and the civil sector to provide leadership on climate. State and local actions would have to become the dominant actors in 20th-century climate policy. When Clinton left office in 2001, no federal action had been undertaken on climate change. Policy action moved to subnational governments and actors.

20th-Century U.S. State and Local Actions

State Efforts

Several state efforts emerged in the early years after the federal government failed to ratify the Kyoto Protocol. In 1997 Oregon became the first state to regulate GHG emissions by requiring new power plants to offset approximately 17 percent of their emissions. Power plants were allowed to do this either by improving their efficiency or by purchasing offsets. The offsets included actions taken outside of the power plant operations such as reforestation, landfill gas recovery, or methane capture (Rahm 2010). New Jersey soon followed suit.

New Jersey, in 1998, became the first state to put in place across the economy cuts in GHGs. New Jersey's governor Christine Todd Whitman by executive order set the goal of reducing New Jersey's emissions 3.5 percent below 1990 levels by 2005. (Whitman would later become EPA Administrator in the George W. Bush administration.) New Jersey put in place a coordinated strategy using every state agency and sector to reach its goals. New Jersey also entered into a partnership with the Netherlands which established a system of emission trading between New Jersey and the Netherlands. The state also set up a system of flexibility in permit approvals with private companies if the companies voluntarily agreed to reduce emissions (Rabe 2002).

Local Actions

Localities got involved in the climate change debate and took action largely through the creation of climate action plans (CAPs). The nation's first CAP was put in place in Portland, Oregon, in 1993. While there is no generally agreed upon definition of a CAP, they typically take the form of a jurisdiction-wide roadmap which outlines what a government or agency will do to reduce emissions, or mitigation efforts. Climate action plans frequently also address the issue of how the locality will adapt to living in a warmer world and all that comes with it including local events of heavy precipitation, drought, wildfire, heat waves, storm surge and sea-level rise (for coastal communities). Adaptation at the local level often overlaps with emergency management services.

Climate action plans are statements of policy for energy, transportation, and conservation and generally do not have any legal power until further steps are taken by elected officials to enact zoning or building ordinances to implement the CAP. The first step in creating a CAP is carbon assessment but after the assessment is done urban planners get involved creating land-use code and ordinances and green building ordinances such as cool rooftops code. Equity concerns may also be introduced by such actions as expanding public transit options, lowering transit fees, and opening cooling centers for those who lack access to air conditioning (Bodin 2019).

Many of the early municipal efforts were driven by the International Council for Local Environmental Initiatives (ICLEI), formed in 1990 with the goal of helping local governments achieve international impacts by focusing on local improvements in environmental policy. In 1991, ICLEI moved to address the issue of climate change when it launched its Urban CO_2 Reduction Project in 14 cities in the U.S. and Europe. That project ran until 1993 with the goal of reducing local CO_2

emissions and documenting strategies for such reductions. In 1993, building on the success of the CO^2 Reduction Project, ICLEI created the Cities for Climate Protection (CCP) campaign. The CCP campaign initially sought to enlist cities that produced 10 percent of the world's CO^2 emissions and to show them how to reduce those emissions. By 2001, CCP had more than 400 members worldwide with 79 in the U.S.

Those localities included: Alachua County, FL; Albuquerque, NM; Ann Arbor, MI; Arcata, CA; Arlington, MA; Aspen, CO; Atlanta, GA; Austin, TX; Berkeley, CA; Boston, MA; Boulder, CO; Bridgeport, CT; Brookline, MA; Broward County, FL; Burien, WA; Burlington, VT; Cambridge, MA; Chapel Hill, NC; Charleston, SC; Chicago, IL; Chittenden County, VT; Chula Vista, CA; Corvallis, OR; Dane County, FL; Davis, CA; Decatur, GA; Delta County, MI; Denver, CO; Durham, NC; Fairfax, VA; Fort Collins, CO; Hillsborough County, FL; Honolulu, HI; Keene, NH; Little Rock, AR; Los Angeles, CA; Louisville, KY; Lynn, MA; Madison, WI; Maplewood, NJ; Medford, MA; Memphis, TN; Mesa, AZ; Miami Beach, FL; Miami-Dade County, FL; Milwaukee, WI; Minneapolis, MN; Missoula, MT; Montgomery County, MD; Mount Rainier, MD; New Orleans, LA; New York, NY; Newark, NJ; Newton, MA; Oakland, CA; Olympia, WA; Orange County, FL; Overland Park, KS; Philadelphia, PA; Portland, OR; Prince George's County, MD; Rivera Beach, FL; Sacramento, CA; Saint Paul, MN; Salt Lake City, UT; San Diego; CA; San Francisco, CA; San Jose, CA; Santa Cruz, CA; Santa Fe, NM; Santa Monica, CA; Schenectady County, NY; Seattle, WA; Springfield, MA; Tacoma Park, MD; Tampa, FL; Toledo, OH; Tucson, AZ; and West Hollywood, CA (Betsill 2001).

To become a member of CCP, the locality had to pass a local resolution showing the locality's willingness to address the threat of climate change by reducing GHG emissions. Localities also had to agree to participate in a series of steps to address climate change including baseline assessment, setting a reduction target, and preparing an implementation plan. CCP members received technical assistance from ICLEI and some funding from the Environmental Protection Agency (Betsill 2001).

The Rise of Opposition

Part of the legacy that the 21st century would inherit from earlier times was the organized resistance of many stakeholders to taking action to stop climate change. Some of the prominent actors in this movement were international while others focused on U.S. politics. The international actors include the world's oil industry led by Exxon-

Mobil. All of the fossil fuel sector was part of the opposition but in the latter part of the 20th-century coal and natural gas were less internationally-focused industries and so took a back seat to big oil.

In 1989, ExxonMobil and the American Petroleum Institute formed the Global Climate Coalition. This coalition was also joined by other energy, automotive, and industrial companies. The goal of the Global Climate Coalition was to oppose any policy action on climate change. ExxonMobil and the Coalition argued that global warming was a natural phenomenon and that human activities were not contributing to it. The strategy drew on tactics pioneered by the tobacco industry in the 1960s to deny that tobacco products were addictive or caused disease. Part of the tactic was to promote doubt and uncertainty. For climate change, the approach that was adopted was to promote doubt in the public mind that climate change was caused by the burning of fossil fuel (Shulman 2007).

With the signing of the Kyoto Protocol in 1997, Shell, BP, and Texaco reversed position, accepted the scientific consensus on climate change, and withdrew from the Coalition. In 1999 Ford Motor Company also withdrew (Bradsher 1999). These dropouts marked the beginning of the decline of the Coalition, which eventually disbanded in 2002 (Gerwin 2002). But ExxonMobil was not yet done. In 1998, ExxonMobil put together a task force called the Global Climate Science Team (GCST). Its members included Randy Randol, ExxonMobil's chief environmental lobbyist; Joe Walker, the American Petroleum Institute's public relations staffer; and Steven Milloy, the head of Advancement of Sound Science Coalition, a nonprofit originally created by Phillip Morris in 1993 to raise doubts about the negative effects of secondhand smoke. The GCST created a Global Climate Science Communications Action Plan in 1998. A six-page memo forwarded to the team by Joe Walker summarized the action plan. In it, he incorrectly stated that there was no scientific evidence of climate change actually occurring and, if it was occurring, whether humans have any influence on it (Walker 1998). The memo went on to introduce a strategy to follow which was to spread misinformation to the media, the public, and members of Congress to sow doubt about climate science.

In subsequent years, ExxonMobil followed its action plan closely. It did shift its approach somewhat to support independent third-party groups that also endorsed its position for not regulating emissions. Between 1998 and 2004 ExxonMobil spent over $16 million to fund organizations that focused on creating public uncertainty about the scientific consensus on climate change. Among the third-party groups ExxonMobil worked with were anti-regulatory groups like the conservative American Enterprise Institute and the Cato Institute. Later ExxonMobil widened its outreach to include other lesser-known groups including the American Council for Capital Formation Center for Policy Research, the George C. Marshall Institute, the American Legislative Exchange Council, and the Committee for a Constructive Tomorrow (Shulman 2007).

Many of the third parties that ExxonMobil funded were small overlapping groups of individuals, some of them scientists, who discounted the consensus in scientific literature on climate change. Many of these groups had the same board members and scientific advisors. Among these groups were the Competitive Enterprise Institute originally founded in 1984 to advance libertarian ideas but later moved to anti-climate issues. ExxonMobil also funded the George Marshall Institute, founded in 1984, and a prominent leader in climate denialism. In 1998, the George Marshall Institute partnered with the Oregon Institute of Science and Medicine in a petition drive sent to thousands of scientists across the U.S. The petition mailing included an article that was formatted to look like one that might appear in the National Academy of Sciences (NAS), a prestigious peer-reviewed scientific publication. The article discounted the scientific consensus on climate change; however, the article had not been peer-reviewed nor accepted for publication in NAS, which later released a statement disavowing any connection to the petition or the article (Shulman 2007).

Other groups also formed portions of the opposition. These included anti-science conservatives and deregulatory Republicans. Deregulation was a Republican mantra beginning with the election of Ronald Reagan in 1980. Reagan also targeted the EPA in unprecedented ways up to that date in time. Environmentalism had been generally bipartisan in the U.S. from its early political beginnings in the late 1960s but with the election of Reagan that bipartisan consensus began to erode. The deregulatory push focused on several sectors of the economy including transportation and utilities, but environmental regulation was also targeted. The problem of climate change and the approach taken in the Kyoto Protocol pointed to the need for more, not less, regulation. This led to a backlash among many Republicans. While Reagan's successor to the White House George H.W. Bush was an advocate of climate action, many Republicans broke with him on his approach. By the time Clinton had assumed office, more and more Republicans had adopted the deregulatory agenda (Rahm 2010).

Opposition to the Kyoto Protocol was driven mainly by economic concerns but it was also underpinned by the deregulatory mindset of many Republicans. Many studies have shown that conservatism is negatively related to pro-environmental attitudes because environmental protections typically rely on some form of government action that restricts economic libertarianism. The conservative movement also saw efforts to restrict GHG emissions as particularly threatening because it was based on a binding international treaty that they felt threatened national sovereignty (McCright and Dunlap, Defeating Kyoto: The Conservative Movement's Impact on U.S. Climate Change Policy 2003).

In an analysis of 14 conservative think tanks active in the 1990s, McCright and Dunlap analyzed hundreds of documents and publications pertaining to climate change and found that nearly three-quarters of them tried to discredit scientific evidence. They emphasized the uncertainty of climate science, describing it as contradictory, flawed, murky, and junk science. The think tanks denied that there was a scientific consensus on climate change, and they often denied the fact of a warming globe. They attacked the scientists as having a hidden agenda. They claimed that the IPCC was a political rather than a scientific organization (McCright and Dunlap, Challenging Global Warming as a Social Problem: An Analysis of the Conservative Movement's Counter-Claims 2000).

Another group that gained prominence in the 1980s and 1990s was the fundamentalist Christian conservative movement under the auspices of the so-called Moral Majority movement. They opposed abortion and stem cell research. They differed from mainstream Christians on the role of science and the role it played in their faith. Former President Jimmy Carter, an evangelical Christian, framed a major difference between evangelicals and fundamentalists as what could be learned from science. Carter, like many mainstream Christians, saw no conflict between science and religion. Fundamentalists, however, by accepting the literal truth of the Bible, reject scientific teachings which they find in conflict with the literal words of the Bible (Carter 2005).

The split goes back to the 1925 Scopes case which challenged the Tennessee's Butler Act, which made it illegal to teach in public schools any theory that discounted the biblical explanation of creation. The case marked the beginning of the split between the teaching of evolution, which underpins modern life sciences, and the literal story of creation depicted in the Bible. After the Scopes case, fundamentalists retreated to their traditional silence on politics, but the rise of Jerry Falwell's Moral Majority movement changed that. Under Falwell, political lobbying became central to the fundamentalist agenda. The Moral Majority was succeeded in 1989 by the Christian Coalition led by Pat Robertson who continued the emphasis on politics to achieve the Coalition's goals. These included ending legal abortion, opposition to stem cell research, protections for the "traditional family," opposition to LBGTQ rights, and a demand for teaching a viable alternative to evolution in the schools (creationism, creation science, intelligent design). Because this group possessed a general suspicion of science, it was also a likely group to accept the challenges to climate science (Lambright 2008).

By the turn of the 21st century, these groups were active in opposing any policy action on climate change. This opposition was based on different reasons including not wanting to see corporate profits and business models diminished, conservative economic considerations, deregulatory political leanings, anti-international perspectives, as well as moral and religious foundations. They would prove to be a powerful coalition of forces for the years going forward.

Conclusion

Climate activists of the 21st century inherited much from earlier times. The first important matter was the rapidly maturing science of climate change and the role that anthropogenic releases of carbon dioxide, primarily from the burning of fossil fuels, and other GHGs played in the warming of the planet. That science began in the 1800s and rapidly expanded in the 1900s. By the turn of the 21st century, much of the fundamental understanding of the science of climate change was well underway and a general consensus among scientists had emerged: climate change was the result of human actions.

With this new awareness, a worldwide political movement emerged to take up the challenge. Beginning with the creation of several UN special groups to address the issue, climate concerns rose to be important enough for the passage of the UNFCCC and later the Kyoto Protocol. Unfortunately, U.S. politics shifted between the presidency of George H.W. Bush, who supported action, and Bill Clinton, who also supported action but was unable to rally Congressional support.

The 20th century ended with a growing climate change opposition movement championed by ExxonMobil and other corporate leaders who feared for loss of profits if the movement to mitigate GHGs won. Joining the opposition was a group of politically conservative activists that, for a variety of reasons including religion, economics, anti-internationalism, and deregulatory politics, opposed taking action on climate change. This was the setting when George W. Bush took office, becoming the first president of the 21st century.

CHAPTER 4

Eight Years of Policy Turbulence

The George W. Bush Administration

Introduction

The 21st century ushered in the new administration of George W. Bush as the 43rd president of the United States. Bush served two terms from 2001 to 2009. Over these eight years many developments occurred on the issue of climate change. Key among them was the release of major new Intergovernmental Panel on Climate Change (IPCC) reports and assessments which strengthened the science on climate change. Despite the ever better science, political opposition in the U.S. grew as was shown with Bush's rejection of the Kyoto Protocol soon after coming into office. During the Bush years there was continued pressure brought by stakeholders to deny the science of climate change and to oppose any policy action. This was largely done through the influence of the fossil fuel sector backed by other powerful actors.

Also, during these years the world saw the international implementation of the Kyoto Protocol. The failure of the U.S. national government to take any policy action on climate change moved domestic policy to subnational governments in the U.S. A wide array of regional, state, and local policies were implemented in an attempt to move the U.S. in a direction that would be in step with other nations. Activists pushed back on federal inaction by using the courts to push policy forward. *Massachusetts v. EPA*, decided in 2007, began a new era of subnational government pressure on the federal government for meaningful policy action on climate change. This chapter details the events of Bush's two terms in office.

The First Term: 2001–2004

George W. Bush came to office under the shadow of a contested election. Voting differentials in many counties were close, leaving the final vote in question. In 2000, the U.S. Supreme Court halted the Florida recount, throwing the election to Bush. His opponent, Democratic candidate Al Gore, conceded the election, making George. W. Bush the president-elect (University of Virginia Miller Center 2022).

Bush's Withdrawal from the Kyoto Protocol

Quickly after becoming president, Bush announced on March 29 his intention to abandon ratification of the Kyoto Protocol. In a March 2001 press release from the White House, reprinted in *Energy & Environment*, President Bush explained his reasons for opposing the Kyoto Protocol. His explanation hinged on several factors. First, he said, "I oppose the Kyoto Protocol because it exempts 80 percent of the world, including major population centers such as China and India, from compliance, and would cause serious harm to the US economy." Later in the press release he also stated that carbon dioxide (CO_2) was not a pollutant listed in the Clean Air Act. In addition, he made the argument that coal provides for half of U.S. electric generation and imposing restrictions on CO_2 would result in shifts away from coal and much higher electricity prices. Finally, he argued that any policy that could harm consumers needed to be carefully evaluated given the "incomplete state of scientific knowledge and the causes of, and solutions to, global climate change..." (G. W. Bush 2001).

The withdrawal of the U.S. from the Kyoto Protocol had profound results. While other countries would move forward, the fact that the U.S., then the world's largest emitter of greenhouse gases (GHGs), pulled out gave other nations' profound concern that no matter their efforts, climate change would not be seriously addressed. One thing that the withdrawal crippled was the emerging carbon market. By reducing participants in the pool of traders, the power of the former Soviet Union was enlarged. Because the Russian Federation was initially given too many credits, with participation of the U.S. in the system, the Russians were able to flood the market with what came to be called hot air trading. This hot air trading largely undermined the emissions trading scheme originally adopted by the Kyoto Protocol, thus massively decreasing the environmental effectiveness of the treaty (Bohringer and Loschel 2003).

In addition to sabotaging the flexibility mechanisms that the U.S. had fought to be included in the treaty, U.S. withdrawal also damaged the image of the U.S. worldwide. Pacific Islanders, the European Union, Japan, and Australia all condemned the withdrawal as irresponsible. The claim was made by national leaders that the Kyoto Protocol simply would not work without U.S. participation. The withdrawal called into question American world leadership and increased the likelihood that other nations would withdraw (Christie 2001).

Despite U.S. withdrawal, the Kyoto Protocol was ratified by the necessary number of countries and went into effect on February 16, 2005, when the Russian Federation submitted its instrument of ratification. The following year the Clean Development Mechanism opened. In 2006, at the 13th COP in Bali, attendees adopted the Bali Road Map and action plan which charted the course for adoption of a new plan to reduce CO^2 emissions. The plan included aspects of mitigation, adaptation, technology, and financing. In 2008, Joint Implementation started. In the month before Bush left office, delegates met again at Copenhagen to begin negotiations on a post–Kyoto agreement (UNFCCC 2020b). What came out of the Copenhagen meeting, though, was generally characterized as a failure since it did not produce a binding treaty, instead merely extending the end date of the Kyoto Protocol to 2020 (Ottinger 2010).

Anti-Science Initiatives

In line with his conservative ideology, Bush changed federal government funding policy on stem cell research. Research on exiting lines would be allowed but extraction of new stem cells from embryos was prohibited. Later in his administration Bush would also move to ban late-term abortions by signing into law newly passed legislation (University of Virginia Miller Center 2022). This stance was in accord with the more conservative religious base that had supported Bush and portended a general anti-science direction of the administration.

Dick Cheney's Energy Task Force

President Bush created the Energy Task Force in 2001 via executive order of January 29. Officially called the National Energy Policy Development Group, the group produced a National Energy Policy Report in May of 2001. The objective of the task force, which was chaired by Vice President Dick Cheney, was to define an energy policy for the country. The U.S. had experienced energy turmoil in 2000–2001 caused primarily by a scarcity of oil. The U.S. had lost its oil production capability and had become increasingly dependent on imported oil to meet domestic needs. At that time, petroleum was considered the most critical source of fuel for the country as it composed two-thirds of the energy demand. Energy turmoil had prompted President Bush to establish the Energy

Task Force and to seek a recommendation on national energy policy. There were two directions the task force could go. It could recommend that the country stay on the same path and continue to consume oil at a rate that created a great import dependency, or it could have recommended a shift of policy to renewables and conservation, gradually reducing petroleum use. The task force considered these issues and released its report on May 17, 2001 (Klare 2004).

While the group was supposed to be composed of government officials, media reports alleged that it was attended by lobbyists and leaders of the fossil fuel industry, including the CEO of Enron, a large energy broker, Kenneth Lay. The environmental group National Resources Defense Council tried to get records of meeting notes but was denied. The group Judicial Watch and the Sierra Club filed suits to obtain copies of the attendee's names. By the time environmental groups were called in to talk to the task force, Cheney and the group had already held more than 40 meetings with fossil fuel energy lobbyists and other fossil fuel interests who supported policies of drilling in the Arctic National Wildlife Refuge (ANWR) and reducing environmental regulations on power plants.

Secret Service records later showed that the meetings had been attended by representatives of BP, Shell, Enron, the American Petroleum Institute, the National Mining Association, the U.S. Oil and Gas Association, Conoco, and ExxonMobil. Among other things, they pushed to have the Department of Energy (DOE), rather than the Environmental Protection Agency (EPA), have authority over energy policy, believing that the DOE would be far friendlier to their activities (Sourcewatch 2020). In 2005 a court ruling finally declared that the White House did not have to release its records to the public. This determination was made on the basis of the claim that the task force was composed of government employees who are not required to be identified (Reporters Committee for Freedom of the Press 2005).

The report was called the National Energy Policy (NEP) and while it initially seemed to favor increased use of renewables and conservation, it actually supported increased use of petroleum domestically and increased production within the U.S. by supporting exploration and drilling in wilderness areas including ANWR. The NEP also favored securing more reliable sources of foreign oil for continued importation. It recommended ways to secure oil from oil-rich regions including the Persian Gulf, the Caspian Sea region, West Africa, and Latin America. It linked Bush energy policy to military policy. It determined not to redirect energy policy to conservation and renewables but rather to stay on the path of continued oil consumption (Klare 2004).

By embracing the NEP, the Bush administration also made a commitment against concern with climate change. This crucial decision to continue on the path of increasing use of fossil fuels, and petroleum in particular, set the stage for other aspects of Bush's policies including climate change policy.

Terrorism Takes Center Stage

The terrorists' attacks on New York City and Arlington, VA, on September 11, 2001, became one of the defining events of the Bush presidency. The administration's response to these attacks was first to go to war in Afghanistan, which under Taliban control had provided sanctuary for Al Qaeda and its leader, Osama bin Laden. Bush then turned his attention to Saddam Hussein of Iraq. In his 2002 State of the Union address, Bush called attention to "the axis of evil" which included North Korea, Iran, and Iraq. In September of 2002, Bush began seeking domestic political support for an attack on Iraq and got bipartisan support in the form of an authorization for the use of force in October of 2002; and in 2003 Bush told the public that the U.S. was at war with Iraq. Despite the debunking of Bush's claims about weapons of mass destruction in Iraq, Bush and his oilman Vice President Dick Cheney were renominated as the Republican candidates in 2004 and Bush was reelected in November of 2004 (University of Virginia Miller Center 2022).

Energy Intensity Goals

Bush announced modest targets for GHG intensity reductions in 2002 by calling for a national goal to reduce energy intensity by 18 percent by 2012 (White House 2008). Energy intensity is a measure of energy consumed per unit of output or energy consumption and gross domestic product (GDP). Historically, economic growth led to more energy consumption. The measure of energy intensity indicates if there is a decoupling of energy use and economic growth. When energy consumption grows slower than the economy, this is an indicator of decoupling (European Environmental Agency 2019). By using this measure, the Bush administration was able to paint a rosier picture of U.S. emissions than would be the case if absolute amounts of GHG emissions were used.

Bipartisan Bills in the Senate 2003

Despite the Bush rejection of the Kyoto Protocol there was still an effort mounted by a group of senators to tackle the climate problem. In the 108th Congress (2003–2004), a bipartisan effort to do something about climate change took form. Led by Senator John McCain (R–Ariz.) and Senator Joe Lieberman (D–Conn.) The Climate Stewardship Act of 2003 was introduced. It put in place a cap-and-trade system to reduce emissions from several sectors of the economy including electricity, manufacturing, commercial, and transportation. Together these represented 85 percent of the economy. The bill had a House companion that was introduced in the 109th and 110th Congresses (Center for Climate and Energy Solutions n.d.). The bill initially did not receive a vote, so Lieberman introduced it again the following year; however, it did not pass.

The Second Term: 2005–2009

Bush was inaugurated January 20, 2005, and sworn in for his second term. Iraq and Afghanistan were still key focuses of the administration's efforts. Climate change remained an important issue for many in the country and efforts continued to urge the Bush administration to do more to address the issue. Bush, however, pushed back on these efforts and remained a supporter of continued use of fossil fuels and even the expansion of the fossil fuel industry. But the Bush administration did have to deal with the ever increasing dependence on imported oil.

Energy Policy Act of 2005

The Energy Policy Act of 2005 was an attempt to deal with the increasing dependence on imported oil. One of the main features of the act was the establishment of a national renewable fuel standard that mandated a vastly increased use of biofuels. The law established $14.5 billion in incentives to be spent over 11 years. Specifically, the law created $1.3 billion in incentives for conservation and energy efficiency, $4.5 billion for renewable energy, $2.6 billion for oil and gas production, $3 billion for coal production, and $3 billion for electricity generation. The law required that beginning in 2006 the nation's gasoline had to contain at least 4 billion gallons of ethanol or biodiesel and that this amount was to increase by 700 million gallons annually through 2012.

The law contained provisions to increase production of gas and oil on federal lands and was supported by fossil fuel advocates including Joe Barton (R–Texas) who defended the fossil fuel provisions of the law by arguing that to lower dependence on imported oil the country needed to spur domestic production. Opponents of the law, however, argued that the measure gave little to conservation and renewables and much more to traditional oil and gas interests (Ballotopedia nd).

Sowing Seeds of Scientific Doubt in Federal Agencies

The large number of deregulatory Republicans and fundamentalist Christians active during the Bush years created pressure for the administration to use doubt of science to deny or delay any policy action on climate change. Within operations of the administration these efforts came to the surface in different ways. One was the editing of EPA scientific reports about climate change to include statements of doubt or uncertainty of scientific findings. In addition to EPA, the National Aeronautics and Space Administration (NASA) scientist James Hansen's reports and public comments were also edited to emphasize uncertainty about the science. Hansen, however, fought back, publicly complaining about the editing. He made headlines when an employee in NASA's Public Affairs Office threatened Hansen with payback should he continue his public statements. Hansen was not deterred, and he took his complaints about the politicization of science to the press. When it was discovered that the public affairs official had lied about having a university degree on his resume, he was fired. Then NASA established internal procedures that stated that NASA scientists could speak to the public and the press only on issues of science but if they spoke on policy, they needed to do so as private citizens (Lambright 2008).

The Bush administration took other actions to undermine Hansen and other NASA scientists. For instance, they formally changed NASA's mission statement to remove the phrase "to understand and protect our home planet" and in doing so, altered the direction of research within the agency. Scientists at NASA argued that without the phrase it would be far more difficult to pursue science that focused on climate change (Revkin, NASA's Goals Delete Mention of Home Planet 2006).

The administration also took aim at the Fish and Wildlife Service (FWS) by censoring biologists and other scientists working in the Arctic over the issue of whether the polar bear should be put on the Endangered Species List. The FWS scientists were told they could not discuss climate change, polar bears, or sea ice unless they were specifically authorized to do so (Revkin 2007a).

In 2007, the Office of Management and Budget (OMB) cut the written testimony of the head of the Centers for Disease Control and Prevention (CDC) to a Senate committee investigating the health risks associated with climate change. The language cut had drawn the conclusion that climate change posed serious health risks (Revkin, Climate Change Testimony Edited by White House 2007).

Anti-Science Initiatives

In 2006, the Congress passed a bill to remove restrictions on stem cell research, but Bush issued the first veto of his administration to kill the bill, remaining aligned with conservative positions on stem cells. There were insufficient votes in the House to override the veto. Bush's argument was that the bill would have taken innocent life in the hope of finding scientific cures for diseases. Attending the White House ceremony were children who were adopted as frozen embryos left at fertility clinics and later born (Bash and Walsh 2006). In June of 2007, Bush issued his second veto of a measure seeking to lift scientific research restrictions on human embryos. The move put Bush at odds with the general public's opinion on such research (Stolberg 2007).

Energy Independence and Security Act of 2007

The Energy Independence and Security Act (EISA) was passed by Congress and signed into law by President Bush in 2007. The law had several key features that extended features of the Energy Policy Act of 2005. In particular, the law expanded the production of renewable energy by setting a higher standard for mandatory use of biofuels, raised the corporate average fuel economy (CAFE) standards to 35 miles per gallon by 2020, to provide for energy efficient lighting and other appliances, and to make buildings more energy efficient (IEA 2017). What the bill did not do was to remove fossil fuel tax subsidies.

Mandated Emissions Reporting

In 2007, Congress passed the Fiscal Year 2008 Consolidated Appropriations Act which contained a provision that EPA was ordered to publish a rule requiring public reporting of GHG emissions from large sources, thus creating the Greenhouse Gas Reporting Program database. This database provides nationwide GHG emissions data, thus supplementing the reporting already then being done by the electric sector under the Clean Air Act Amendments of 1990. This provision added over 40 source groups to the reporting program (Center for Climate and Energy Solutions n.d.).

The Bush Administration and the Courts

A series of court cases were decided during the Bush years that had important consequences for climate change. One of these battles was between the state of California and the federal government. California had held a special status in air pollution control since the 1960s. California had passed a state air pollution control law before the federal government passed the Clean Air Act. When the federal statute was passed, states were banned from passing separate state laws governing vehicle emissions, except for California. The federal law had created an exemption for California, which was required to seek a waiver of preemption from the EPA when it wanted to impose a higher standard. The Clean Air Act was also later amended to allow other states to adopt California's standard but only after California received a waiver (California Air Resources Board 2019).

In 2002 California passed a law that placed limits on one source of GHGs—automobiles. The bill allowed the California Air Resources Board (CARB) to set limits on tailpipe emissions by requiring that all cars and light duty trucks sold in California beginning with model year 2009 have 22 percent decreased emissions of CO^2 compared to 2002. These emissions were further set to decline by 30 percent by model year 2016 (Holtcamp 2007). Since 2002, eleven states followed California's standards and set GHG standards for automobiles. California, Vermont, and Rhode Island were all sued by the automotive industry to block these requirements. A fourth case was brought against New York for its attempts to control GHG emissions from new power plants. Each of these cases stalled in the courts as they waited for a decision on yet another case that had come under U.S. Supreme Court review (Fialka 2007).

One of the most important court cases of the Bush years was *Massachusetts v. EPA.* The ruling was handed down on April 2, 2007. The court ruled in the case that EPA had the authority under the Clean Air Act to regulate GHGs in new motor vehicles. In addition, the court said that EPA could not avoid its responsibility to regulate GHG emissions unless it could show the scientific basis on which to do so (Greenhouse 2007). The case had been brought by a group of states including California, Connecticut, Illinois, Maine, Massachusetts, New Jersey, New Mexico, Oregon, Rhode Island, Vermont, and Washington; local government including the District of Columbia, New York City, and Baltimore; and a group of private interests including the Center for Biological Diversity, the Center for Food Safety, the Conservation Law Foundation, Environmental Defense, Friends of the Earth, Greenpeace, the National Environmental Trust, Natural Resources Defense Council, the Sierra Club, the Union of Concerned Scientists, and U.S. Public Interest Research Group (Massachusetts v. Environmental Protection Agency 2007). These groups had committed to GHG reductions. But even after the court ruled on the case, the Bush administration dragged its feet. The EPA issued a finding of endangerment in 2009 before the end of the administration and still the EPA did not move to bring GHGs under regulation.

After the ruling, California asked EPA to take action on its 2005 waiver request but again the EPA stalled, and California ended up taking EPA to court. The other cases that were pending also began to move through the courts. The district courts ruled against automakers in 2007, saying Vermont did have the authority to regulate GHG emissions coming from vehicles. The courts also ruled that California could regulate emissions. But none of these cases could be enforced unless EPA granted California the waiver (Egelko 2007). The EPA denied the waiver in December of 2007, saying that the newly passed Energy Independence and Security Act of 2007 had raised CAFE standards to 35mpg making California's waiver unnecessary because the primary way California would reduce emissions was by raising fuel economy. California and 15 other states filed suit against the Bush administration for the denial of the waiver in 2008 (Roosevelt 2008).

International Implementation of the Kyoto Protocol

The Kyoto Protocol came into effect in 2005, after the release of the IPCC Fourth Assessment Report, which increasingly warned of the challenge that climate change posted. It imposed binding reduction targets for developed nations and a set of innovative policy mechanisms to assist in reaching those reductions. Among these innovative policy mechanisms was the establishment of a carbon market and carbon trading, joint implementation (JI), and the clean development mechanism (CDM). The Kyoto Protocol allowed parties to add or subtract from their initially assigned target amounts through land use, land use change and forestry (LULUCF) activities and through participation in the Kyoto mechanisms of emission trading, JI and CDM.

Together these three mechanisms were established to allow Annex I countries to take advantage of lower-cost emission reduction strategies outside of their territory. Parties were allowed to set up country- or regional- based trading systems. One example was the EU Emissions

Trading Scheme (EU ETS). Joint implementation was established as a project-based system that allowed one Annex I nation to undertake an emissions reduction project in another Annex I nation. The CDM allowed an Annex I nation to undertake emissions reduction projects in developing countries (de Boer 2008).

Details of the mechanisms had to be negotiated prior to their coming into effect. This happened at the sixth and seventh COPs in Bonn and Marrakesh in 2001. The Marrakesh Accords set the stage for the ratification of the Kyoto Protocol with formal approval of the operating rules for emissions trading, JI, and CDM along with compliance rules and accounting procedures. Emission Trading launched in 2005 with the creation of the EU ETS and in that year, with the ratification of the agreement by the Russian Federation, the Kyoto Protocol came into force. In 2006, the CDM began. In 2007, in the 13th COP in Bali, the Bali Action Plan was adopted, marking the beginning of more extensive planning to combat climate change. The JI mechanism began in 2008 (UNFCCC 2020b).

The IPCC During the Bush Years

A number of IPCC reports appeared during the Bush years. Beginning in January of 2001, just as Bush was taking office, the Third series of assessment reports was released. The TAR Climate Change 2001: The Scientific Basis. Also released in 2001 was TAR Climate Change 2001: Impacts, Adaptation, and Vulnerabilities; TAR Climate Change 2001: Mitigation; and the TAR Climate Change 2001: Synthesis Report. These report built on past reports but included information collected between 1996 and 2001 which resulted in a better understanding of science. The reports indicated that human activities continued to drive climate change. In March of 2005, the IPCC released two special reports. One was on Carbon Capture and Storage and the other was on Safeguarding the Ozone Layer and the Global Climate System. The fourth assessment reports (AR4) of the IPCC were released in 2007 and the series included the separate working group reports of The Scientific Basis; Impacts, Adaptation, and Vulnerabilities; Mitigation; and the Synthesis (IPCC, History of the IPCC 2022).

The change in tone between the scientific reports released at the beginning of the Bush administration and those released in 2007 was marked. The 2007 reports for the first time said that global warming is "unequivocal" and that human activity is "very likely" driving most of the warming. The AR4 for the first time concluded with more than 90 percent confidence (near certainty) that human actions since 1950 were the main cause of warming. The report also stated that future warming was certain if behavior did not change (INFORSE 2007). Despite the growing scientific consensus on climate change, opposition to policy action grew.

The Role of the Opposition During the Bush Years

Over the course of the Bush administration, the partisan divide over climate change grew with Republicans becoming increasingly opposed to both the science of climate change and any role the government might have in taking policy action. This can be seen by Gallup polling which annually asked a series of questions and catalogued results by political affiliation. Respondents were asked to self-identify their political affiliation. In response to the question, "Do you think the effects of global warming have already begun to happen?" Republicans in 2001 answered affirmatively 48 percent of the time while Democrats answered yes 61 percent of the time. By 2008 the gap had increased with only 41 percent of Republicans but 76 percent of Democrats answering in the affirmative.

The percent saying that global warming is exaggerated by the press went up for Republicans, rising from 43 percent in 2001 to 59 percent in 2008. For Democrats, 18 percent said the press exaggerated global warming consistently for the entire period. In response to the assertion that most scientists believe global warming is occurring, 56 percent of Republicans responded positively and that percentage remained consistent throughout the period. Democratic support for the assertion went from 70 percent in 2001 to 74 percent in 2008.

In 2003 Gallup introduced another question, asking if Earth's changing temperatures over the last century were due more to human activities than to natural changes in the environment. In 2003, 52 percent of Republicans and 68 percent of Democrats agreed. By 2008, only 42 percent of Republican agreed while 73 percent of Democrats agreed. The percentage saying global warming will pose a serious threat to them and their way of life in their lifetimes went from 21 percent of Republicans in 2001 to 29 percent in 2008. For Democrats the percentages shifted from 41 percent in 2001 to 50 percent in 2008 (Dunlap 2008).

Republican elected officials were more extreme than the general members of the population that affiliated as Republicans. In preparing Republicans for the 2004 race, a Republican advisor, Frank Luntz, released a memo with some suggested language that politicians could use while running for 2004 elections. Luntz explained to readers that Republicans were vulnerable on environmental issues and in the lengthy memo he suggested tactics Republicans could take to soften their position on environmental issues with the public. Specifically on climate change, Luntz wrote, "voters believe there is *no consensus* about global warming within the scientific community." Should the public come to believe that the scientific issues are settled, their views about global warming will change accordingly. "Therefore, you need to make the lack of scientific certainty a primary issue in the debate...." Luntz went on later in the memo to say "the scientific debate is closing [against us] but not yet closed. There is still a window of opportunity to challenge the science" (Luntz 2003).

By making this argument, Luntz was admitting the science was clear that policy actions would need to be taken to combat climate change but because Republicans held deregulatory positions and received much funding from the oil, gas, and coal industries, the strategy Luntz proposed was to deny science and delay action as long as possible.

The combination of deregulatory Republicans and Christian fundamentalists in the Bush administration's base of supporter moved the Republican party to doubt science and delay any policy action as long as possible. As discussed above, one way to do this was to introduce doubt of science into the regulatory agencies, as the Bush administration did with EPA, NASA, CDC and FWS by editing reports and restricting government scientists talking points. Another was by restructuring agencies like NASA by altering their missions to make focusing on climate change more difficult (Rahm 2010).

Oil companies were extremely effective in lobbying for their position on climate change policy during the Bush years. In addition to becoming inside advisors to the Cheney Energy Task Force, Exxon-Mobil continued its relentless misinformation campaign and spent considerable sums of money in its efforts. A 2007 Union of Concerned Scientists report traced many third-party anti–climate change organizations to ExxonMobil funding. These groups included the Competitive Enterprise Institute (CEI), the George C. Marshall Institute, and the Committee for a Constructive Tomorrow. ExxonMobil funded many of these third-party groups with more than $1 million per year to support their anti-science, anti-global warming efforts (Shulman 2007).

The media also played an important role in the climate change debate in large part by relying on a position of fair and equal treatment of both sides. But the role played by the media was not fair or equal. The media instead would give equal time to climate skeptics or deniers thus giving the impression that they were half of the opinion on the matter. What really would have been the fair position would have been to give 90 percent of the attention to the scientific consensus and very little airtime to the opposition. In framing the issue in the way they did, the media did not rely on science but rather on opinion.

Climate science was pitted against anti-science climate deniers or skeptics, thus giving the opposition more credibility than they deserved. The media frequently gave equal time to voices who simply said they did not believe in climate change, as if climate science was a matter of belief rather than fact and empirical observation (Boykoff and Boykoff 2007). The rise of the Internet during this time period also played a role as conservative web sites, largely unpoliced, hosted unabashed misinformation. Some of these sites were supported by industry. These sites largely went unchallenged until 2004 when NASA scientist Gavin Schmidt, speaking as an individual citizen under NASA's newly adopted rules for scientific conduct, launched a site supporting the science of climate change (Leslie 2004).

Although the role of the media during the Bush years overwhelmingly supported the side of the opposition, by 2004 the media was beginning to shift. At this time a number of movies, television dramatizations, and weather events showed that the media could play a more balanced role. In 2004, the movie *The Day After Tomorrow* was released. Its plot line was the onset of a rapid and severe weather shift that doomed the U.S. population and forced relocation of those surviving the crisis to central America. In 2007, Al Gore's *An Inconvenient Truth* played in theaters. It won the Academy Award for best documentary in 2007 and was watched by millions. It depicted climate change in a way that was accessible to ordinary people. The companion book hit number 1 on the *New York Times* best seller list.

Scientists were split on Gore's message, with some saying Gore went too far in depicting scientific certainty and others rejecting Gore's assertion that Hurricane Katrina could be tied to climate change. At the time the science of event attribution was in its infancy and most scientists did not go so far as to link individual weather events to climate change (Broad 2007). The opposition used the debate to once again cast doubt on the science behind climate change. While scientists might have been hesitant to link individual events to climate change, the public was less hesitant. Hurricane Katrina had received a great deal of national coverage, and to many in the public it appeared to be proof of what scientists had been warning (Weiner 2007).

Regional, State and Local Policy Action During the Bush Years

With national policy stalled in the Bush years, much of the advocacy action moved to subnational governments, the civil sector, and a growing number of private sector organizations willing to take on the issue under the name of sustainability. Regional, state, and local governments played a significant role during this time. States and localities had taken some actions prior to Bush moving to the White House, but after Bush's withdrawal from the Kyoto Protocol, regional, state, and local activism increased.

The first regional agreement, the Climate Action Plan, was established in 2001 by several northeastern states including Maine, New Hampshire, Vermont, Massachusetts, Rhode Island, and Connecticut and the Canadian provinces of Nova Scotia, Newfoundland and Labrador, Prince Edward Island, New Brunswick, and Quebec. The parties agreed to lower GHG emissions to 1990 levels by 2010 and to 10 percent below 1990 levels by 2020. The plan was based on "no regrets" measures, that is, measures that would both reduce energy costs and GHGs at the same time, accordingly the plan emphasized energy efficiency. It also called for creation of a standardized GHG inventory, increasing public awareness, and investigating an emissions trading scheme (Selin and Vandeveer 2005).

To implement their obligations under the Climate Action Plan, some of the participants set up the Regional Greenhouse Gas Initiative (RGGI), joined initially by Connecticut, Delaware, Maine, New Hampshire, New Jersey, New York, and Vermont. The RGGI established the first U.S. cap-and-trade program for power plant emissions. It went into effect December 20, 2005, under a memorandum of understanding signed by the governors of these states. Maryland, Massachusetts, and Rhode Island joined in 2007. Emissions were initially capped at 2009 levels and subsequently reduced by 10 percent by 2019.

The RGGI began its first auction of CO^2 emissions in 2008, so it was fully functional before Bush left the White House. In the first base period from 2006 to 2008, RGGI electric generation plants reduced GHG emissions by 48 percent. States agreed to accept individual shares of the overall RGGI commitment based on negotiations among the states and historical emissions. States were then required to implement state regulations based on a Model Rule developed by RGGI. The Model Rule was adopted in 2008 and modified in 2013 after a 2012 program review (Center for Climate and Energy Solutions 2022).

In 2003, Oregon, Washington, and Oregon joined the West Coast Governors' Global Warming Initiative. The initiative was the response by these governors to Bush's inaction and they pledged coordinated action by the states to fight global warming. The actions they said they would pursue including partnering with businesses to adopt new or improved technologies. In 2006, this alliance was linked to RGGI (NRDC 2003).

Population growth in the West was one of the greatest drivers of energy consumption. To deal with this issue, in 2004 the Western Governors' Association (WGA) adopted the Clean and Diversified Energy Initiative. One purpose of this initiative was to explore the feasibility by 2015 of generating 30,000 megawatts of energy from wind, solar, geothermal, combined heat and power, biomass, and hydroelectric. A second goal was to improve energy efficiency by 20 percent by 2020, and to create a reliable transmission grid to support these new sources of power. While this initiative did not directly address climate change, reduction of energy use in fact would result in a reduction of GHGs (Rahm 2010).

To track the success of adding renewables to the energy portfolios of the states, the West developed the Western Renewable Energy Generation Information System (WREGIS), which began operations in 2007. It allowed for renewable energy certificate (REC) trading across the state members (WECC 2015). In 2007, the Western Climate Initiative (WCI) was created by the governors of Arizona, California, New Mexico, Oregon, and Washington. The WCI was created to promote mitigation strategies using market-based emissions trading (WCI 2013).

In the Midwest an alliance was formed in 2007 called Powering the Plains (PTP). This was a voluntary regional agreement among renewable energy advocates, agricultural interests, states, and industry partners with the goal of finding ways to combat climate change while not impacting the economy. North and South Dakota, Iowa, Minnesota, Wisconsin, and the Canadian province of Manitoba were members. This alliance eventually led to a formal agreement by the states, the Midwestern Greenhouse Gas Reduction Accord (MGGRA) (Crabtree 2008). In 2007 the governors of Illinois, Iowa, Kansas, Michigan, Minnesota, Wisconsin, and the Canadian province of Manitoba signed the MGGRA. By signing this agreement, the states pledged to set GHG reduction targets within one year, to establish a cap-and-trade system to achieve those reductions, and to join the Climate Change Registry to track those emissions (DOE 2007).

Local action was also important. In 2005, the Mayors Climate Protection Agreement was launched by Seattle Mayor Greg Nickels. The goal was to advance the goals of the Kyoto Protocol through leadership and action in American cities. By the time of the meeting of the 2005 U.S. Conference of Mayors annual meeting, 141 U.S. cities had joined. By 2007, more than 500 cities had joined. Under the agreement mayors pledge to have their cities meet or beat Kyoto Protocol targets through local land use policies and public awareness campaigns, and to put pressure on members of Congress and the federal government to enact policies to meet or beat the Kyoto Protocols 7 percent reduction targets (United States Conference of Mayors 2021).

During the Bush years a number of partnerships between the civil sector, business, and governments emerged to deal with climate change. For instance, in 2008, the Nature Conservancy created the Southwest Climate Change Initiative (SWCCI). This initiative was created to engage conservation and land managers in local-scale climate change planning and action. The initiative was aimed at the Four Corners region (Arizona, Colorado, New Mexico, and Utah). It sought to apply a vulnerability assessment tool developed by the U.S. Forest Service, and to apply an adaptation planning scheme developed by the Wildlife Conservation Society (WCS) and the National Center for Ecological Analyses and Synthesis (NCEAS). The project led to further cooperation among the Nature Conservancy, Western Water Assessment, and Southwest Climate Change Initiative (created in 2006 by Arizona and New Mexico). It also led to collaboration with the National Oceanic and Atmospheric Administration (NOAA) (NOAA 2022).

Conclusion

During the two terms of the Bush administration, many policy shifts impacted climate change policy both domestically and internationally. Bush began his first term with the dramatic withdrawal from the Kyoto Protocol. The failure of the U.S. to participate in the accord made the agreement far less effective since the emissions-trading flexibility mechanisms were sabotaged. Nevertheless, the treaty went into force during the Bush years. The isolation of the U.S. from the path other nations were accepting derailed U.S. leadership and created considerable criticism. The treaty was implemented, without the U.S. And the IPCC made considerable progress in releasing major reports on the science of climate change that made the case for policy action more dire.

The Bush administration was openly hostile to science—both the science that underpinned climate change but also the science that was held in contempt by many of his conservative and fundamentalist supporter. The Bush administration, through censorship and editing of scientific reports, introduced a weapon mastered by the fossil fuel industry—doubt—into government deliberations. The partisan divide over taking policy action on climate change would grow during these years in part because that doubt spread through Republican ranks. As the years passed, the powerful industries that would financially gain from business as usual poured millions into efforts to undermine any attempts to forge a national policy on the climate crisis.

Energy was a large focus of the Bush years. Part of the emphasis on energy resulted from the terrorist attacks on the U.S. in 2001 and the subsequent wars launched by Bush in response to those attacks. The war in Afghanistan had some direct link to the attacks but the war in Iraq was largely underpinned by Bush's later debunked claims of weapons of mass destruction. The war in Iraq came to be seen by opponents of the war as about oil. These wars, and the continued threat of terrorism, spurred many in the U.S. to reconsider energy security and how to obtain it. Rather than taking the path toward shifting away from fossil fuels, Bush's Energy Task Force led by Dick Cheney set the U.S. on the road to more oil importation and greater use of fossil fuels.

During part of the Bush years there were some signs of the potential to move forward on climate change policy. During Bush's first term there was some continued bipartisan movement, at least in the Senate, to develop climate policies in the form of a cap-and-trade bill. But those efforts failed. Nevertheless, Bush was forced to at least address the issue of climate change, but he did so using a measure, energy intensity, that masked the real problem. The Energy Policy Act of 2005 and the Energy Security Act of 2007 did introduce some incentives to advance renewables but did not withdraw the massive subsidies for oil, gas, and coal built into the system. Mandated emissions reporting did move forward. The battles in the courts eventually resulted in a win for climate activists, especially with the results in *Massachusetts v. EPA*. But the administration stalled on implementing the remedies mandated by the court.

With no viable federal action, policy in the U.S. fell to subnational governments. Regional, state, and local actions were successful in creating a number of coalitions, alliances, and agreements to fight climate change. During the Bush years, this is where meaningful climate policy was made.

CHAPTER 5

Eight Years
of Attempted Progress

The Barack H. Obama Administration

Introduction

The administration of Barack Obama put the U.S. back on a course to address the climate crisis. Obama had stressed during his campaign a desire to do something about climate change, however, events of the first term including the continuing financial crisis that had begun in 2008 and the Democratic Party's deep desire to pass legislation on healthcare reform often took priority. During the years of the Obama administration, the country experienced a series of mass shooting events which also created pressure to do something about gun control. With these competing issues, combating climate change often took a second place role in policy formulation.

During the first term, the Obama administration was able to use the financial crisis to forge climate policy by investing funds from the American Recovery and Reinvestment Act in green energy and energy efficiency. There was also an attempt to pass climate legislation. When the legislative attempt failed, and upon losing the House in the midterm elections, the administration decided to adopt an executive approach to climate actions.

After winning a second term in 2012, the Obama administration moved to take action on climate change, albeit through executive actions. The Clean Power Plan was the administration's key idea to move climate action forward, however, this plan was immediately challenged in the courts and was never implemented. Obama's second term domestically was marked by legal fights to overcome opposition to his actions but internationally the president made some headway. Traveling to China in 2014, Obama was able to win a commitment from Xi Jinping to cut emissions. This commitment was important as it ended the prior approach that let developing countries avoid any emissions reductions. Obama also made headway at the Lima climate summit when the idea of Nationally Determined Contributions replaced the outdated bifurcated system of developed versus developing countries. This led the way to the successful negotiation of the Paris Agreement in 2015.

Even though the science on climate change made significant advances in the Obama years, political opposition to action grew. This was largely because the Republican Party moved in a hard right populist direction that vehemently rejected climate science. With campaign funding unlimited, conservative financial support to oppose climate action was massive. Industry giants like ExxonMobil also heavily funded opposition efforts. Despite this organized opposition, regional, state and local action surged forward. This chapter discusses in some detail the major efforts to move climate policy ahead during the Obama years and the forces that pushed back on those efforts.

The First Term: 2009–2012

Barack Obama was inaugurated on January 20, 2009, after winning the November 2008 election against his Republican opponent Sen. John McCain (R-AZ). The immediate issues that confronted the administration were financial. The economy had entered into financial crisis at the end of the Bush administration and early efforts of the new administration focused on passing a stimulus bill to help the economy recover. The Obama administration was able to use that opportunity to steer funds into renewable energy and energy efficiency, thus partially fulfilling a campaign pledge to address the climate crisis. The other events that would dominate the first term included attempts to pass climate legislation, passage of the Affordable Care Act (ObamaCare), efforts to create a more just country for the LGBTQ community, a number of public shooter tragedies resulting in Democratic attempts to pass gun control measures, and the assassination of Osama bin Laden (the mastermind of the 9-11 attacks on the U.S.).

Copenhagen COP15

Before being sworn in as president, president-elect Obama traveled to Copenhagen to attend the 15th Conference of Parties (COP15) of the United Nations Framework Convention on Climate Change (UNFCCC). The meeting, scheduled for December 7–18, 2009, sought to negotiate a successor treaty to the Kyoto Protocol, however, negotiations were unsuccessful. Rather, the meeting resulted in political commitments from all major economies—including China for the first time—to cut emissions. The basic terms of the agreement at Copenhagen were negotiated personally by the president-elect and a few leaders of developing countries on the last day of the conference. It took another full day of bitter negotiations to arrive at a procedural compromise to allow the arrangement made by the leaders to be formally accepted. This occurred with harsh objection from several developing countries. The agreement did not include a statement of legally binding commitments. The agreement did include an aspirational goal of limiting temperature increases to 2 degrees Celsius, a process for countries to submit their emission pledges by 2010, terms for reporting and verification, and a commitment of funds to help developing countries reach their targets (Center for Climate & Energy Solutions 2009).

When Obama was sworn into office in January of 2009, the main issue facing the nation was the financial crisis which had begun in 2008. The Bush administration had put in place some measures to deal with the crisis, including the Troubled Assets Relief Program (TARP) which enabled the federal government to pump money into the banking and mortgage system to avoid financial collapse. The TARP was made possible by the passage of the Emergency Economic Stabilization Act of 2008 (History 2018). While this measure was somewhat successful, more needed to be done to regain financial stability. Under the Obama administration, these efforts continued with the passage of the American Recovery and Reinvestment Act (ARRA).

The American Recovery and Reinvestment Act of 2009

The ARRA was primarily concerned with repairing the economy from the financial crisis of 2008, but it also contained several provisions that made it an energy bill of considerable importance. The historic $787 billion bill devoted more than $45 billion to energy efficiency and alternative energy programs. It included $13 billion to make public housing and federal buildings more energy efficient and to weatherize more than a million residential homes. It included $20 billion for further development of renewable energy power, $2 billion for research on carbon capture and storage, and $18 billion for a wide variety of environmental programs. The bill also provided tax credits of up to $7,500 for the purchase of plug-in hybrid cars (Broder 2009). It gave the Department of Energy (DOE) $4.5 billion to modernize the electric power grid and DOE subsequently established the Small Grid Investment Grant (SGIG) program and partnered with 200 participating electric utilities to put in place 99 projects to modernize the grid. Many of these focused on smart grid technologies (DOE 2022). The smart grid program centered on using advanced information and communications technologies (ICT) to enable much wider use of renewables by making them easier to integrate into the grid (Ekanayake et al. 2012). By including these energy provisions, the ARRA was the administration's first step at climate policy.

Attempts to Pass Climate Legislation

In his first speech to Congress, in a joint session held on February 14, 2009, President Obama asked Congress to send him a climate bill with a cap-and-trade provision for carbon emissions (Rahm 2010). In June, the House passed the climate bill and sent it to the Senate (James 2009). The Waxman-Markey Bill, or the American Clean Energy and Security Act, contained five key provisions covering clean energy, energy efficiency, reducing global warming pollution, transitioning to a clean energy economy, and agriculture and forestry. The primary item in the bill was for the creation of a cap-and-trade system for carbon dioxide (CO_2), methane (CH^4), nitrous oxide (N^2O), hydrofluorocarbons (HFCs), perfluorocarbons (PFCs), sulfur hexafluoride (SF^6), and nitrogen trifluoride (NF^3). Regulated entities were to include refineries, importers of petroleum fuels, and distributors of natural gas. The bill also established provisions to control black carbon. The bill set caps to reduce greenhouse gas (GHG) emissions by 3 percent below 2005 amounts by 2012, rising to 83 percent by 2050 (Center for Climate and Energy Solutions 2009b).

The bill never came the floor for discussion in the Senate. The failure in the Senate had several causes. First, the persistent economic crisis made senators more susceptible to big oil, gas, and coal interests who lobbied hard against the bill. Those special interests made the claim that the bill would further hurt the economy. While many in the Senate supported the bill, there was insufficient support to overcome the 60-vote threshold of the filibuster. By this point in time the Republicans had grown increasingly hostile to the science of global warming and to the idea of cap-and-trade, seeing it more as a tax on energy than an innovative market-based solution. Oil, gas, and coal interests spent more than $500 million lobbying against the bill between January of 2009 and June of 2010. ExxonMobil was the number one spender (Weiss 2010).

Dominance over Congress played an important role in President

Obama's ability to achieve his agenda and how to try to achieve it. While the Democrats held majorities in the Senate throughout the first term of Obama's administration, they did not have a filibuster-proof super-majority for most of the time. The Republican minority leader of the Senate, Mitch McConnell (R-KY), was less than cooperative in working with Democrats. Indeed, in 2010 Mitch McConnell said that his main goal was to assure that Obama be a one-term president (Kessler 2017). In the mid-term elections in 2010, the Democrats held the Senate but lost the House. Moreover, the Democrats still did not have enough Senators to overcome a filibuster.

The Second Term 2013–2017

In November of 2012, Obama won re-election. During that election cycle, the Democrats continued to hold the Senate, albeit without a filibuster-proof supermajority, but the Republicans held the House. This power arrangement assured the continued lack of ability to use legislative approaches to achieving policy goals. Events also drove the agenda.

In December of 2012, there was another major mass shooting event. At Sandy Hook elementary school 20 children and 6 adults were killed. With this event, the beginning of Obama's second term was marked again by his call for gun control, a call that was rejected by the Senate. Distractions from climate also followed in 2013 from the revelations of Edward Snowden regarding National Security Agency (NSA) domestic surveillance. But a major speech given at Georgetown University in June of 2013 returned the focus to climate issues.

New Climate Action Plan

On June 25, 2013, Obama spoke at Georgetown on climate change. Obama said he had chosen the location because youth clearly would be more affected than others by the adverse effects of climate change. After briefly talking about the science of climate change, the President stated that he believed it was time to act and announced a new national climate action plan which included a call on the Environmental Protection Agency (EPA) to end carbon pollution by regulating power plants. Another feature of the action plan was to double energy from wind and solar. To do so, the President instructed the Department of the Interior (DOI) to authorize enough wind and solar on federal land to accomplish the goal. The third goal announced in the speech was that the federal government would consume no less than 20 percent of its energy from renewables by 2020 (Obama 2013).

Because of the difficulty of pursuing a legislation-based strategy to achieve policy goals, the Democrats moved to an executive action strategy. In his 2014 State of the Union Address, President Obama pledged a year of executive action to bypass Congress. A number of executive actions were taken during 2014 and after in an attempt to achieve his agenda without having to go through Congress. For instance, in February President Obama ordered the EPA to begin work on the next phase of fuel-efficiency standards to increase the mileage of heavy-duty vehicles and to propose new incentives for medium- and heavy-duty trucks to run on alternative fuels, and in September, Obama announced a series of public-private partnerships to increase the use of solar technologies and to promote energy efficiency.

In May of 2014, the President announced a series of executive actions to reduce carbon pollution, to prepare the U.S. for adaptation efforts, and to participate in international efforts to fight climate change. On May 6, 2014, the U.S. released its third U.S. National Climate Assessment, which emphasized the fact that climate change was already affecting the U.S. At the 2014 UN Climate Summit, President Obama announced an executive order in Climate Resilient International Development which required all U.S. agencies to factor climate resilience into all international development work. In November of 2014, the President announced the U.S.'s intention to contribute $3 billion to the Climate Resilience Fund to help reduce carbon emissions in developing countries (White House of President Obama 2014).

The Clean Power Plan

One of the most consequential of the executive actions taken in 2014 was the announcement and release in June of the Clean Power Plan (CPP). The plan proposed carbon reduction strategies for domestic power plants that would put the nation on a path to reducing carbon emissions from power plants by 32 percent by 2030. The EPA, under the authority given it by the Clean Air Act, established standards to reduce carbon emissions from new and existing power plants. Under the rule, the EPA established performance standards that new and existing power plants must meet. The Clean Power Plan allowed states to determine their own plans to meet these standards. Each state was required to submit a plan to the EPA indicating how it would meet the performance standards specified in the Clean Power Plan given the mix of electricity generating units in the state (Federal Register 2015).

Almost immediately after the publication of the final rule that put the Clean Power Plan into force, a coalition of 24 states led by West Virginia and Texas and one coal company (Murry Energy) filed a lawsuit arguing that the EPA was exceeding its lawful authority and asked the U.S. Court of Appeals for the District of Columbia Circuit to immediately stop implementation of the rule until its legality could be determined. The EPA rule was supported by 18 states (including California and New York) and many environmental organizations. To block the immediate execution of the new standards the opposition coalition was required to show the court that immediate implementation of the rule, while the court challenge to the rule went forward through normal court processes, would cause irreparable harm.

Irreparable harm was exactly what those opposition states said would happen. They argued that EPA's rule, in essence, was an illegal demand that would force the reorganization of the nation's electricity grids. They argued that this attempt to restructure such a large part of the nation's economy was beyond the power Congress granted EPA under the Clean Air Act. They also said that the rule's "fence line" provisions fell outside of EPA's authority. By this they meant that under the rule the EPA would allow states to reach their targeted emission levels not only by reducing power plant emissions but also by adding renewables and other carbon-neutral sources of electric generation. The last two would be outside the "fence line" of the power plant and therefore, according to opponents, outside of EPA's legal jurisdiction.

The EPA and its supporters, however, argued that no immediate irreparable harm would result because under the rule the states would have as many as three years to develop their plans and that the real implementation date was seven years in the future. Nevertheless, in February 2016, the U.S. Supreme Court issued a temporary halt to the CPP, stopping states from having to comply with the rule until the host of lawsuits could be resolved (Magill 2016). By the end of the Obama administration, the CPP issue had not been resolved.

2014 UN Climate Summit

In September of 2014, President Obama traveled to New York to give a speech and participate in the Climate Summit. The Climate Summit was a meeting of heads of state brought together by the Secretary-General of the United Nations, Ban Ki-moon. The Summit was intended to build momentum for the Paris meeting to be held a year later. At this summit, both the U.S. and China called for a binding agreement to emerge from Paris. At the summit, the Secretary-General called for green growth in developing countries by scaling up renewable energy projects. Saving forests was also a large topic of conversation as was sustainable agriculture. The summit was accompanied by a large march that was joined by over 400,000. The march pointed to the increasing public demand for climate action (Henry 2014).

U.S.-China Joint Agreement

In a trip to Beijing in November of 2014, President Obama jointly announced with China's President Xi Jinping an agreement to cut CO^2 emissions. The U.S. set a target to reduce GHG emissions by 26–28 percent by 2025 while China announced its intention to increase the share of non-fossil fuels in its energy mix to around 20 percent and to peak emissions by 2030 (White House of President Obama 2014). While the mid-term election resulted in the Republicans holding the House and picking up the Senate, the Obama administration argued that the U.S. could make these commitments without Congressional approval. Nevertheless, Republicans were critical of the arrangement, saying that China had gotten a free pass (Echeverria and Gass 2014).

Lima COP20

In the effort to move forward with a climate negotiations in Paris in 2015, attendees of the Lima Conference of Parties, held in December of 2014, agreed to bring their "intended nationally determined contributions" to the Paris negotiations. The Lima meeting was the third of a four year round of meetings that was scheduled to end in Paris in 2015. It was contentious in that developing countries continued in large part to insist on common but differentiated responsibilities but by substituting the intended nationally determined contributions (INDC) approach, Lima largely sidestepped the issue. However, disagreement persisted on what exactly should be in the INDCs. Developed countries wanted the INDCs to focus only on mitigation, while developing countries insisted on including adaptation and finance, as well. The final agreement linked the INDCs to the final objective of the UNFCCC, to stabilize GHG emissions to avoid dangerous anthropogenic interference with the climate system, thus forcing mitigation efforts as a priority. However, the agreement also invited parties to consider adaptation measures as well (Center for Climate & Energy Solutions 2014).

Rejection of Keystone XL Pipeline

On November 6, 2015, the Obama administration rejected the Keystone XL Pipeline. The pipeline was to connect Canadian tar sands oil fields to American refineries. The rejection by Obama was based on climate change. The administration announced that America was again a global leader on climate action and was rejecting the pipeline because of the impact the combustion of such dirty oil would have on climate change (Goldenberg and Roberts 2015). The Keystone XL pipeline, however, was not permanently killed. President Obama's successor in the White House would revive the decade-long battle over the pipeline.

The Paris Agreement

The Paris Agreement, the international accord replacing the Kyoto Protocol, was agreed to in Paris on December 12, 2015, by 195 nations. The Paris Agreement elaborated, in 31 pages, several major goals and ways to achieve them. The text called the Paris Agreement was actually two documents, the Paris Decision and the Paris Agreement. The Paris Decision section, composed of 140 separate decisions, listed "Decisions to Give Effect to the Agreement" and was not legally binding. The Paris Agreement section listed the 29 Articles of the legally binding agreement. The three primary goals were specified in Article 2 of the agreement. Perhaps the most important goal set was to limit the increase in global average temperature to "well below 2°C (3.6°F) above pre-industrial levels" and to try to limit that increase to 1.5 °C. The second goal set was to help countries adapt to living in a warmer world. The third goal was to provide money so that poorer countries could pursue "climate resilient development." How to achieve these goals was specified in more detail in the agreement (UNFCCC 2015).

All countries were required under the agreement to prepare an assessment called a nationally determined contribution (NDC). The initial INDCs were the starting point of the Paris negotiations as determined by the Lima COP20 in 2014. However, preliminary analysis of the initial round of INDCs suggested that if all nations reached their stated goals, global average temperature would rise by 2.7°C, which was more than the goal stated in Article 2. So, the agreement required that all nations prepare a succession of NDCs, with each new NDC being an improvement over the last one submitted. The NDCs were to be submitted in five year intervals. The inclusion of all nations—both developed and developing—into the Paris Agreement through these voluntary national plans was an entirely new way to approach international climate talks. While the Paris Agreement still acknowledged that rich nations should proceed with more haste in reducing emissions than poor nations, nevertheless, all nations were to be held accountable for their actions (UNFCCC 2015).

All nations were required to "conserve and enhance" all resources that act to absorb greenhouse gases, especially forests. The agreement allowed use of the REDD+ process: Reducing Emissions from Deforestation and Forest Degradation with the + to include conservation, sustainable management, and enhancements of forests. The agreement specifically allowed parties to fund reforestation efforts in developing countries and get credits (offsets) toward meeting their targeted goals in exchange. The agreement also supported an alternative approach preferred by some, called Joint Mitigation and Adaptation (JMA), which argued that by giving indigenous people strong legal rights to their land, their communities would be strengthened, and, in turn, they would protect the forests.

The agreement acknowledged that the poor nations needed financial assistance if they were to reduce their emissions and adapt to a warmer world while continuing to work their way out of poverty. The Paris Agreement section said funding was expected to be "continuous and enhanced" over time while the decision section acknowledged that $100 billion would be needed annually. Under the agreement, rich nations were expected to report, biennially, how their financing efforts were progressing. In addition to direct finance, the rich nations were expected to transfer technologies that might assist developing nations with adaptation and mitigation efforts, as well as to provide developing nations with capacity building to achieve these goals (UNFCCC 2015).

While no outside agency was set up to assure that each nation met its commitments, all nations were required to account for their NDCs using transparent, accurate, and comparable measurement methods. All nations were required to report national anthropogenic greenhouse gas inventories using the good practices established by the Intergovernmental Panel on Climate Change. Nations were required to report both sources of gases and removals (sinks). These reports were to be done in such a way that they show progress in achieving the NDCs.

Finally, periodically the parties to the agreement agreed to meet and take stock of their progress (called the "global stocktake"). The first global stocktake was scheduled to occur in 2023 and every five years thereafter. The results of the global stocktake were to be used by all parties to revise their NDCs so that the world could make its commitment to limit temperature increases to "well below 2° C" if not all the way to 1.5° C. However, item 20 of the decision section called for an earlier "facilitative dialogue" to take place in 2018 and at that meeting countries were also supposed to update and improve their NDCs (UNFCCC 2015).

Marrakech Climate Talks

On November 7, 2016, international talks began again on exploring and improving decisions made in Paris. The talks lasted until November 18 as world leaders tried to come to accord on three provisions of the Paris Agreement: transparency, finances, and how to improve outcomes by getting countries to increase their greenhouse gas reduction pledges. The discussion on transparency focused on how countries would be held accountable for reporting and reaching their NDCs. The objective of the talks regarding finance was to create a mechanism whereby the richer nations could help the poorer nations adapt to a warming world and to reduce their GHG emissions. The final point under discussion centered on the fact that even if every country reached their originally promised NDC reductions, the reduction in GHGs would still not be sufficient to meet the warming goal established in Paris. That stated goal was to keep warming at no more than 2 degrees Celsius but to try to limit warming to considerably less than 2 degrees Celsius.

The Paris Agreement went into force November 4, 2016. The Marrakech talks were the first held to implement the agreement. Much of the talks ended up celebrating the fact that the world adopted the Paris Agreement so rapidly. The talks produced the Marrakech Action Proclamation which was a strong statement of commitment to moving forward with the implementation of the Paris Agreement (UNFCCC 2016). The conference also produced the Marrakech Partnership for Global Climate Action, a path to accelerating climate action between 2016 and 2020 by supporting public-private partnerships (UNFCCC 2016). Despite these agreements, final decisions on finance and transparency were not forthcoming. Delegates to the talks were thrown into a bit of disarray when, on the second day of the conference, Donald Trump became president-elect of the United States.

Intergovernmental Panel on Climate Change (IPCC) Reports During the Obama Administration

The IPCC released a number of important scientific reports during the years of the Obama administration. The first appeared in 2011. It was a special report on renewable energy sources and climate change mitigation. It included updates on the newest advancements in renewable energy, including specifics on geothermal, wind, bioenergy, solar, hydroelectric, and ocean energy. Also included in the report were analyses of the integration of renewables into the existing grid and future energy systems, renewable energy in the context of sustainable development, mitigation potential and costs, and policy, financing, and implementation. A 2012 report focused on managing the risks of extreme events and disasters to advance climate change adaptation.

By 2013, the IPCC had begun to release its Fifth Assessment Report (AR5) beginning with the volume on the Physical Basis which was a comprehensive assessment of the science of climate change since the release of the AR4 in 2007. It included observations about the cryosphere (those portions of the Earth's surface where water is in solid form), the atmosphere and surface, and the oceans. It also contained special sections on clouds and aerosols, carbon and other biogeochemical cycles, anthropogenic and natural radiative forcing, climate models, detection and attribution of climate change from global to regional, near- and long-term climate change projections, and sea-level change.

In the summary for policy makers, the report stated that "warming of the climate system is unequivocal" and that "the atmosphere and oceans have warmed, the amounts of snow and ice have diminished, sea level has risen, and the concentration of greenhouse gases has increased" (IPCC 2013).

Two other reports in 2014 completed the AR5 and they included a volume on mitigation and a synthesis report. The synthesis report opened with the overall assessment stating "Human influence on the climate is clear, and recent anthropogenic emissions of greenhouse gases are the highest in history. Recent climate changes have had widespread impacts on human and natural systems" (IPCC 2014). All together the AR5 painted an increasingly deteriorating situation with dire consequences lacking sufficient political action to reduce emissions. Yet, despite these dire warnings, the power and influence of the opposition grew during the Obama administration.

The Opposition
During the Obama Administration

Misinformation was one of the largest sources of opposition during the Obama administration. Between 1997 and 2018 ExxonMobil spent $37 million to spread climate misinformation, much of it during the

Obama years. During that same period of time, conservative petrochemical magnates Charles and David Koch spent $145 million to fund climate-change-denying think tanks and advocacy groups. In the 1990s fossil fuel interests had given rise to the Global Climate Coalition (GCC) to push back on efforts to reduce GHG emissions and to fight the Kyoto Protocol. They did so largely by promoting doubt of the climate science.

In the 1990s, they funded a small group of scientists and pseudoscientists willing to take money to deny the science of climate change. These scientists got a lot of publicity in the media because at the time the media was presenting the view that there were two legitimate sides of the climate science debate, which was incorrect. Giving equal time to a small number of outliers was hardly balanced and the mainstream media later changed its approach to emphasize the scientific consensus. But during the debate on the Kyoto Protocol, ExxonMobil ran full page ads in *The New York Times* questioning the science of climate change.

When Obama took office in 2009, a new push to do something about climate change emerged. It took the form of the Waxman-Markey bill. As discussed earlier in this chapter, the bill never passed the Senate. Much of the opposition to the bill came from the American Petroleum Institute (API). With this defeat, the Obama administration moved to regulations to push its climate agenda. These regulations also came under fire, as discussed earlier (Pierre and Neuman 2021).

While the GCC, ExxonMobil, the American Petroleum Institute, Americans for Prosperity, and the Koch brothers remained active during the years of the Obama administration promoting climate denialism, one of their prior allies made some changes. The George C. Marshall Institute officially closed in 2015, although it was really only replaced by a new organization called the CO_2 Coalition. The George C. Marshall Institute had been an active conservative think tank in the U.S. since the 1980s. One of its primary agenda items was to sow seeds of climate denialism (Library of Congress 2015). The stated goal of the new organization, charted as a nonprofit educational organization, was to provide information to "leaders, policy makers and the public about the important contribution made by carbon dioxide to our lives and the economy."

Additionally, they emphasized the "limitations of climate models, and the consequences of mandated reductions in CO_2 emissions." Moreover, the group posted so-called facts on their site that contradicted IPCC reports, including such items as that the planet was experiencing a 140-million-year dangerous decrease in CO_2 levels, that the warming effect of each CO_2 molecule declines as concentrations increase, that the primary function of CO_2 is as a plant food, and that the planet is CO_2

impoverished (CO^2 Coalition 2022). So, while the group had nominally given up its overt climate denial, they still advocated for the so-called constructive role of CO^2 for the economy. Fox News, the conservative Murdock-owed Republican mouthpiece, was listed as one of the media outlets to whom they broadcast their messages.

During the Obama administration, the Republican position of being the party denying climate change and opposing climate policy action grew to orthodoxy. While running against Barack Obama in 2008, the Republican candidate John McCain used a campaign advertisement praising candidate McCain on standing up for climate action. By the end of the Obama years, that position would be dead. The Republican Party shifted over the years from being a party that would debate how best to deal with climate change to a party of full denial. By 2016, the last year of the Obama administration, Donald Trump secured the Republican nomination and proclaimed climate change a Chinese hoax. By 2017, being a climate denier became just another litmus test to prove good Republican credentials. This occurred despite the fact that over the Obama years, the scientific consensus on an anthropogenic cause of climate change grew stronger.

Many of the Republicans were moved in this direction by the Koch brothers and the fact that many of their constituents came from regions dependent on fossil fuel extraction. The turning point was a "No Climate Tax" pledge drafted by the Koch brothers in 2008. The effort gained Republican support after the House passed the Waxman-Markey bill. But the hacking of the climate research program at East Anglia University and misreading and release of scientists' emails, allowed conservatives to argue that scientists were using a statistical "trick" as evidence of climate change.

The scientific research was eventually validated but confirmation came after the Republicans and their climate denying allies in industry and think tanks had done their damage. The fringe in the Republican Party became the mainstream. With the Supreme Court case, *Citizens United*, lifting restrictions on corporate financing, the Koch brothers and Americans for Prosperity began a major campaign to elect lawmakers that would ensure the fossil fuel industry would not have to deal with climate change legislation or regulations. These efforts were largely successful (Davenport and Lipton 2017).

Scholars have debated the reasons for this swing in the U.S. Republican Party (as well as a number of right-wing parties worldwide). They have argued that the swing toward rejection of climate science and policy action was the result of the rise of populism. Populism is a belief that the divide in society is largely between a "pure people" and a "corrupt elite." Populists tend to think the corrupt elite ignores their opinions and views while holding them in contempt. Right-wing populists also point to the role of a disreputable minority, immigrants, as the source of corruption of elites. These movements rapidly manifest as nativist and authoritarian.

Right-wing political groups in the U.S. also have a history of being climate deniers or climate skeptics. This is one reason why there was a major increase in climate skepticism in the U.S. in the 2000s associated with the rise of the "Tea-Party" movement within the Republican Party. The Tea-Party was fueled by the politics of marginalization, where adherents protested that they had become politically and economically sidelined by globalization and technological change. Climate change policies, under this view, were seen as just another action that would worsen the lives of ordinary workers, be they coal miners, farmers, ranchers, or factory workers.

Political leaders in the Republican Party had to attach themselves to this perspective if they wanted support from this base of voters. While the base of the party may cling to this position as justification for rejecting climate action, however, the politicians leading the movement have another motivation. They receive most of their funding from the fossil fuel industries, giving them a separate reason to support climate skepticism and denial. Another linkage between populism and the denial of climate change is owed to the ideology itself. Climate change was seen by right-wing populists as just another idea being pushed by left-wing urban elites out of touch with traditional American values. The other problem was that climate change, being complex, demands complex solutions that the right-wing populist base rebuffs. They argued instead that the cabal of scientists, environmentalists, and left-leaning politicians constituted a corrupt elite that they rejected (Lockwood 2018).

Because of the rise of right-wing populism during the Obama years, the Republican Party began to approach climate change from the perspective that such policies could only be embraced if they benefited the nation and its core people directly. Right-wing populists rejected the role that international organizations might play in addressing the issue of climate change. They also rejected the role of the global scientific community. These beliefs brought into suspicion any efforts on the part of the United Nations or the IPCC, seeing them as inconsistent with nationalistic values. Moreover, the long-term relationship between the Republican Party and the fossil fuel industry, which funded campaigns of climate denial, cemented the relationship. As Donald Trump campaigned in 2016 on the American First ideology, the Republican Party was firmly under the control of right-wing populists (Fiorino 2022).

Another source of opposition during the Obama years was the power of the lobbying groups during the lead up to the Paris Agreement at COP21. As the talks began, in November of 2015, the number of lobbyists that joined the conference to work for producing a meaningful treaty numbered in the tens of thousands. As of February 2015, there were 1,758 observer non-governmental (NGO) organizations admitted by the UN to the Paris meeting. While they did not directly participate in the negotiations along with the official countries' delegates, they played an influential role in contributing to the positions of the parties. Among these NGOs were a large number of business and industry NGOs (called BINGOs), who composed more than 15 percent of the NGOs present, making up one of the largest groups of NGOs at the conference. The role of the BINGOs at the Paris negotiations was primarily to promote flexible market-based approaches, innovation, and technology deployment if they could not successfully derail attempts to regulate emissions.

A main BINGO present was the International Chamber of Commerce, representing the overall interests of the business community. The BINGOs present also represented a number of large emitters (Climate Policy Info Hub 2015). Some of the NGOs involved with the meeting organized a counter-conference to explicitly work against the treaty. Among them were representatives of the Heartland Institute, the Committee for a Constructive Tomorrow, and the Competitive Enterprise Institute. These groups had, for years, worked to preserve coal jobs and stop any significant regulations. But these groups were just a few of the many working to rollback any meaningful climate agreement. Between 2003 and 2010, fossil fuel companies had contributed hundreds of millions of dollars to 91 conservative think tanks involved in the fight against climate change regulations (O'Harro Jr., 2017).

Regional, State, and Local Action

Regional, state, and local action continued under the years of the Obama administration. The Regional Greenhouse Gas Initiative (RGGI) remained active during the Obama years. It functioned to provide a regional budget for GHG emissions for the participating states. The RGGI mandated that all producers of electricity cap their CO_2 emissions. Between 2012 and 2019 they included Connecticut, Delaware, Maine, Maryland, Massachusetts, New Hampshire, New York, Rhode Island, and Vermont. New Jersey had been an earlier participant from 2009 to 2011 but dropped out. Between 2009 and 2017, RGGI's cap for tons of CO_2 fell from 188 million tons to slightly over 84 million tons, effectively lowering the CO_2 emissions considerably (RGGI 2022).

The Western Climate Initiative (WCI) continued to function during the Obama years and grew to include California, Oregon, Washington, Arizona, New Mexico, Utah, and Montana, along with British Columbia, Manitoba, Ontario, and Quebec in Canada. In 2011, a nonprofit organization, WCI, Inc., was formed to support the efforts of state and regional greenhouse trading programs and WCI transitioned into this organization. In 2016, the Governors' Accord for a New Energy Future was signed a group of 17 governors representing all regions of the country. Members of the accord included California, Connecticut, Delaware, Hawaii, Iowa, Massachusetts, Michigan, Minnesota, Nevada, New Hampshire, New York, Oregon, Pennsylvania, Rhode Island, Vermont, Virginia, and Washington. The governors committed to encouraging and installing clean energy, improving energy efficiency, and adopting renewable energy and alternative-fuel vehicles (Center for Climate and Energy Solutions 2022).

The Powering the Plains (PTP) initiative continued during the Obama years, making advances in planning a roadmap for CO_2 reduction. During these years PTP focused on education and climate issues and how the Midwest could commit to clean energy over the long-term (Great Plains Institute 2022).

At the state level, states continued their policy activism into the 2010s. Maine, an early adopter of GHG reductions, extended its targets in the 2010s. Maryland newly adopted a GHG target in 2009 with the goal of reducing emissions to 25 percent below 2006 levels by 2020. In 2016, Maryland extended and increased its target to 40 percent below 2006 levels by 2030. States also tried to limit their emissions through the adoption of Renewable Portfolio Standards (RPS). In 2011, Indiana set a voluntary RPS at a goal of 10 percent by 2025, while South Carolina adopted its RPS in 2014, mandating a 2 percent of energy generation from renewables by 2021. While these states moved forward, some states took steps back. For instance, West Virginia repealed its 2009 RPS, in 2015, after the state legislature went from Democratic to Republican. Oregon strengthened its net metering program in 2016. Climate adaptation efforts moved forward in 2008 and 2009 with Florida, Virginia, Maryland, and California each adopting plans. Each of these states feared sea-level rise, flooding, and storm-surge (Bromley-Trujello and Holman 2020).

Local action continued to be important. The Mayors Climate Protection Agreement continued to function. During the Obama years, the group worked with the federal government to develop the Energy

Efficiency and Conservation Block Grant (EECBG) program. Through this program the federal government transferred $2.8 billion directly to localities for energy efficiency programs (The United States Conference of Mayors 2022).

Conclusion

The administration of Barak Obama came in on a mood of hope, not only because of the "hope poster" used in the campaign but because of the belief by many that the newly elected president might truly do something about the climate crisis confronting the world. The U.S. had contributed the most GHGs to the atmosphere over its history, and the U.S. was still a major producer of GHGs. But the hope would turn quickly to a hard fight.

Unable to get climate legislation passed, the Obama administration was forced to change to executive action. Executive power is immense, but it can and was challenged by those who opposed climate action. The attempt of the administration to use the Clean Power Plan as the blueprint for GHG reductions, and compliance with the Paris Agreement proved a failure. Opponents used the courts to thwart the administration and were successful in their efforts.

Some limited achievements were obtained including the expansion of the clean energy sector through funding embedded in the American Recovery and Reinvestment Act. But during the Obama administration, largely due to the widespread implementation of fracking technology, the U.S. increased its oil and gas production to become the largest producer of oil and gas in the world. The U.S. surpassed Russia in gas production in 2011 and surpassed Saudi Arabia in oil production in 2018. This massive shift in domestic production that occurred during the administration was a game changer not only for the U.S. economy but also for the urgent demand to expand renewable energy. During the Obama administration, the U.S. once again regained its role as a world leader on the issue of climate change but much of that was owed to the presidential role of statesman. The U.S. once again "talked the talk," but finding sufficient domestic allies to implement policies simply did not happen.

In November of 2016, Donald J. Trump was elected president of the United States by taking the electoral college although not the popular vote. Also in that election, the Republicans took the House and the Senate. Trump's election marked the end of eight years of efforts to address climate policy at the federal level. Opposition to climate action had grown massively during the eight years of the Obama administration. The Republican Party had been transformed from the party of John McCain, who supported climate action, to a party in which climate denial had become the norm. And it was on this note that the Obama administration left office.

CHAPTER 6

The Triumph of the Opposition
The Donald J. Trump Administration

Introduction

Donald Trump was elected president in 2016 with an Electoral College win but without receiving the popular vote. He came to office with both the Senate and House with Republican majorities, thus assuring at least initial legislative success. But the day after his inauguration as the 45th President of the United States millions of people around the world participated in the Women's March to show their opposition to the Trump administration and its proposed policies. The Women's March was the largest single-day protest in American history (UVA/Miller Center 2022). The Trump administration was off to a rocky start and would go down in history as one of the most eccentric administrations in American history. Trump would become the only American president to be twice impeached, to refuse to accept his electoral defeat in 2020, and to oppose the peaceful transition of power—a hallmark of American democracy.

Trump's campaign slogan was "Make America Great Again" (MAGA). Trump campaigned on tax cuts for the wealthy and corporations in solidarity with Republican supply-side, trickledown economics. Terrorism was a large issue for the Trump campaign, and he promised to end immigration from terror-prone regions and to introduce extreme vetting of those immigrants allowed to enter the U.S. He promised a ban on Muslims entering the U.S. During his campaign he condoned extreme interrogation methods, including waterboarding. Trump's emphasis on foreign policy during the campaign focused on trade policy. He promised to renegotiate existing trade agreements like the North American Free Trade Agreement (NAFTA) within his first 100 days in office and to identify all foreign trading abuses that harmed U.S. workers and to end them. He targeted the Trans-Pacific Partnership (TPP), a trade agreement among 12 Pacific Rim countries. He labeled China as a currency manipulator.

In the area of healthcare, Trump promised to end the Affordable Care Act (Obamacare). He aligned himself with the National Rifle Association and pledged to protect gun rights. Immigration was a critical issue for the Trump campaign. Trump ran as a "law and order" candidate and was criticized for his rhetoric when he called Mexican immigrants "rapists." He promised to end funding to so-called "sanctuary cities" and to undo all of President Obama's executive actions on immigration including Deferred Action for Childhood Arrivals (DACA). Trump allied with those in education policy that supported school choice. Trump pledged to appoint Supreme Court Justices that would overturn Roe v. Wade, thus ending the constitutional right to abortion in America (NPR 2016).

In terms of the environment, Donald Trump campaigned on making American energy independent by increasing production of domestic oil and gas, as well as by reducing regulations on the sector. He supported fracking. Trump opposed at least one source of renewable energy, saying that wind turbines were an "environmental and aesthetic disaster." He vowed to reinstate the Keystone XL pipeline, which had been rejected by the Obama administration. He promised to rescind Obama's Climate Action Plan and the Waters of the United States rule. During the campaign, his campaign manager Kellyanne Conway said that Trump believed that global warming was naturally occurring and that human activities had nothing to do with it. Trump called climate change a "hoax" and said he would pull the U.S. out of the Paris Agreement (Ballotpedia 2016). Thus, Trump came to office with a substantial anti-environmental agenda.

Governmental Appointments

The team that Trump assembled to be his appointees said much about Trump's overall environmental agenda. Critics argued that the fossil fuel industry had an oversized presence and that most of Trump's appointees had expressed skepticism about climate change science (Sidahmed 2016). Among the most important of Trump's environmental appointments was the Administrator of the Environmental Protection Agency (EPA). Trump appointed Scott Pruitt, the former Attorney General of Oklahoma, who had spent his career as Attorney General by repeatedly suing the EPA.

Pruitt, a zealous deregulator, began his tenure in EPA with a large and vigorous rollback of Obama-era environmental rules. Among them was the Clean Power Plan (CPP) (already tied up in the courts). In 2017, Pruitt had made headlines by questioning the scientific consensus on anthropogenically-caused climate change. In making such statements, Pruitt contradicted his own agency's findings. Such comments did not put Pruitt in jeopardy with the president; on the contrary, they were lauded by the president who also dismissed climate science. Pruitt's term at the EPA was ended with a storm of ethical scandals including more than a dozen federal investigations that finally caused the president to ask for his resignation. Pruitt was replaced by Andrew Wheeler, a former coal lobbyist (Davenport, Friedman and Habberman 2018).

Wheeler was more successful than Pruitt in rolling back environmental protections. Under Wheeler, the EPA sidelined scientists by either disbanding their advisory councils or replacing scientists with industry-friendly participants. Claiming that he wanted to increase the transparency of EPA processes, he severely reduced the number of scientific studies that the EPA could consider in weighing policy. He was successful in gutting the Obama-era coal ash rule. Wheeler recommended unsafe levels of poly- and perfluoroalkyl (PFAS) in drinking water. He proposed changes to the Clean Water Act that would allow wetlands to be further unprotected. He supported weakening the mercury emissions rule. He reversed the Clean Power Plan (Negin 2019).

Trump appointed Ryan Zinke to be Secretary of the Interior, giving him vast control over the nation's public lands. Zinke, like Pruitt, was embroiled in multiple ethics scandals. Zinke's appointment was announced in 2016 and he was confirmed by the Senate in 2017. He left office in 2019 as a result of the barrage of ethics complaints. While in office he did much to harm environmental protections. He restarted coal leasing on public land, leasing that had been ended by the Obama administration. He sped up licensing processes and pushed for leasing agreements at a rapid pace. He rolled back Obama's rule on methane capture. He approved the construction of transmission lines and pipelines on public land. He defended his positions by arguing that U.S. production was more efficient than foreign producers, under the assumption that consumption would not decline. He proposed that decommissioned military bases serve as export terminals for coal. During his tenure as Secretary, the Department of the Interior's handbook's chapter on climate change was deleted (Aton 2018). In 2022, Interior's Inspector General found that Zinke had lied to federal investigators while he was Secretary of the Interior (Kaplan 2022).

David Bernhardt succeeded Ryan Zinke as Secretary of the Interior.

Bernhardt came to office with a history of climate science denial and while in Interior he made it a priority to restrict the discussion of climate change or climate science within the agency. Bernhardt embraced Trump's agenda of expansion of fossil fuels and further expanded leasing options on public land. Prior to serving in Interior, Bernhardt was a lobbyist for the oil and gas industry (Kustin 2019).

Rick Perry was Donald Trump's choice to run the Department of Energy (DOE). In his earlier campaign for the presidency, when a reporter asked him which agencies he would like to target for elimination, Perry comically forgot the name of the Department of Energy. Nevertheless, Perry ran the agency he had previously sworn to eliminate. While Secretary, Perry vigorously promoted liquid natural gas (LNG) as a viable export given the country's fracking boom. He failed to spend the appropriate amount of money on the department's Office of Energy Efficiency & Renewable Energy. He was called before the Senate and berated for the high number of vacancies in that office. While at the same time, he funneled money to programs that supported the most polluting energy sources including coal. Perry resigned after three years in office (Dillion and Brugger 2019).

Trump appointed Rex Tillerson to be Secretary of State, making him the head of an agency charged with negotiations of international treaties, including environmental treaties. Tillerson was the chief executive of one of the world's largest international oil and gas companies, ExxonMobil. Tillerson went to ExxonMobil in 1975 and became CEO in 2006. Tillerson had close ties to Russia's Vladimir Putin, who Tillerson came to know in the 1990s when he ran ExxonMobil's interests in Russia (Forbes 2017). Tillerson would serve only one year in the job of Secretary of State, leaving his post on March 31, 2018. Tillerson was fired by President Trump, an announcement the President made on Twitter before Tillerson could come forward with a statement, revealing some of the chaos already present in the White House.

Tillerson's tenure with State was marked by the retirement or resignation of 60 percent of the State Department's top career-diplomats (Beauchamp 2018). Tillerson was replaced by Michael Pompeo in 2018. Pompeo praised the shrinking ice in the polar regions, arguing it was a major economic breakthrough, despite the well-known effects of climate change. Pompeo also failed to recognize the anthropogenic causes of climate change, falling back on the commonly used Republican comment that the climate always changes (Kelly and Kosinski 2019).

In addition to these, Trump's other appointments shared the perspective of climate denialism or at least skepticism. These included his appointment of Ben Carson to be Secretary of Housing and Urban

Development, Michael Flynn as National Security Advisor, Jeff Sessions as Attorney General, Tim Price as head of Health and Human Services and Wilbur Ross of the Department of Commerce (Sidahmed 2016).

While these appointments of specific personnel were highly significant, Trump went further to create an administrative state full of his loyal followers. In an executive order issued on October 16, 2020, Trump created a new category of federal civil service workers called Schedule F (Trump 2022). This action, if implemented, would have moved thousands of high-level government managers out of their civil service classification which provides protection from firing, into the new Schedule F which strips employees of their protections. Implementation of Schedule F would make it possible for Trump and future presidents to remove thousands of career service merit employees and replace them with political loyalists. Such a move would have been a considerable step back to the spoil system. This executive order was never fully implemented, and no personnel reclassifications took place in the waning days of the Trump administration. President Biden revoked this order on January 22, 2021, with the issuance of Executive Order 14003 (Schulman 2022).

Dismantling Major Climate Policies

Over four years, the Trump administration presided over the rollback of more than 100 environmental rules governing the climate, clean air, water, wildlife, and hazardous substances. Analyses conducted by Harvard Law School and Columbia Law School showed the details of the environmental reversals. Nearly 100 were finalized during the Trump administration and more than a dozen remained in progress at the end of the administration (Popovich, Albeck-Ripka and Pierre-Louis 2021). The major reversals that had an impact on climate policy are detailed below.

Ending the Climate Action Plan

On January 20, 2021, Trump had the White House website updated. The updates showed the new direction of the incoming administration in reversing Obama's Climate Action Plan. The new website stated that "President Trump is committed to eliminating harmful and unnecessary policies such as the Climate Action Plan…" (Reuters 2017). This shift indicated the Trump administration's intention to promote oil, coal, and gas.

Advancing the Keystone XL and Dakota Access Pipelines

On January 24, 2017, Trump issued an executive order reversing Obama's rejection of the Keystone XL pipeline. He also signed an executive order advancing the Dakota Access pipeline (Yang 2017). Both pipelines had been the center of much debate. The Keystone XL was controversial because it was built to carry Canadian "tar sands" oil into American refineries. Environmentalists considered the tar sands to be of great climate threat because they involved the application of a great deal of heat, producing large amounts of GHGs. The Dakota Access pipeline was controversial because it would cross land held by the Standing Rock Sioux Nation. The tribe rejected the pipeline on two accounts. First, they feared it would contaminate their drinking water. Second, they argued it would violate sacred tribal sites. Trump ignored these concerns in favor of the oil and gas industry's desire for better transport options for their products (Yergin 2020).

Ending the Clean Power Plan

In March of 2017, Trump announced that he planned to reverse much of Obama's Clean Power Plan. That effort would come to fulfillment in 2019 supported by EPA Administrator Andrew Wheeler. The rollback of the CPP was a major effort to undo the Obama administration's climate efforts which aimed to cut the nation's greenhouse gas (GHG) emissions from power plants by 32 percent below 2005 levels by 2030. By replacing the CPP rule with a much weaker regulation, the Affordable Clean Energy Rule, Trump's EPA greatly reduced U.S. efforts to meet promised reductions.

The Affordable Clean Energy Rule aimed to reduce power sector emission by between 0.7 percent and 1.5 percent by 2030, a marked reduction from Obama's standards. At the time of the adoption of the new final rule, U.S. emissions were on the rise after years of declines. The Affordable Clean Energy Rule, if fully implemented, would also have raised the number of American deaths from air pollution. Under the Supreme Court ruling in *Massachusetts v. EPA*, the EPA was required to regulate GHG emissions, so the Trump administration could not just eliminate the CPP—it had to replace it with a new rule.

The EPA was required to meet the standard known as "best system of reduction." But under the CPP, the Obama administration included in its systems carbon pricing, switching to cleaner alternative fuels, and capturing carbon dioxide (CO_2) emissions. Under the Affordable

Clean Energy Rule, the EPA included only efficiency as the best system of reduction. By adopting this measure, power plants would only have to produce more energy from the same amount of fuel. The rule was part of an overall effort within the Trump administration to promote the use of coal (Irfan 2019).

The Affordable Clean Energy Rule was contested in the courts and on January 19, 2021, a federal appeals court in the District of Columbia vacated the rule (Institute for Policy Integrity 2022). The issue remained unsettled at the beginning of the Biden administration.

The War on Science

The Trump administration went to extreme measures to undermine scientific integrity. Some of this was owed to Trump's populism rooted in distrust of elites, including scientific experts. Agencies like the EPA have a long history of using scientists as experts to advise on policy. Trump removed EPA's Science Advisory Board, excluded scientists from receiving EPA and other federal grants, and replaced experts with loyalists with less expertise. Trump also tried to impose a new regime whereby health data used for policy analysis had to be completely disclosed to the public. This meant that health and other personal records would become public and was inconsistent with best practice. The ultimate attempt was to eliminate the use of health studies to support regulations for air quality. In climate policy, the undermining of scientific integrity was clearly shown by the manipulation of the social cost of carbon calculation (Fiorino 2022).

Social Cost of Carbon

The social cost of carbon (SCC) is the main economic measure for estimating the costs of climate change. The SCC is based on damages associated with emissions of one additional ton of carbon and is represented in dollars. Conversely it shows the benefits to society for reducing CO_2 emissions by one ton. That number can be compared with mitigation costs. Economic theory suggests that carbon emissions are a negative externality that are best addressed through a Pigouvian tax set equal to the SCC. The federal government began using the SCC calculations for cost-benefit analyses associated with new regulations beginning in the Reagan administration. The SCC was used to value federal tax credits for carbon capture technologies, beginning in 2018. Many states used the SCC as a benchmark against which proposed climate policies might be compared. The SCC also affected the cost of royalties for oil and gas leases on federal land. SCC calculations originated with the work of William Nordhaus in 1982 (Rennert et al. 2021).

On March 28, 2017, President Trump issued an executive order purporting to improve energy independence. The executive order rescinded the Obama-era SCC calculation. At the end of the Obama administration the cost was about $40 per ton of CO^2. Trump ordered a new calculation, one that did not include any impacts except domestic ones. By removing international impacts, the cost of carbon would fall dramatically (Revkin 2017).

Reversal on Coal Mining Ban

In March of 2017, President Trump issued an executive order rolling back the Obama administration's temporary ban on mining coal. But the courts initially ruled that the Trump administration did not include sufficient analysis of the environmental impacts of renewed leasing. After undertaking an environmental assessment, the Bureau of Land Management (BLM) announced in February of 2020 that it had concluded that there would be no significant impacts from rolling back the Obama ban and that it would resume leasing on public lands (Frazen 2020). Coal mining leases on federal land continued throughout Trump's term.

Removing Reporting of Climate Impacts

In an executive order issued March 28, 2017, Trump directed the Council of Environmental Quality (CEQ) to rescind Obama-era guidance on how to account for climate change in National Environmental Policy Act (NEPA) reviews (Reilly 2017); NEPA requires analysis of environmental impacts before committing to a federal project and the CEQ oversees that process. The guidance issued by CEQ told federal agencies exactly how to factor GHGs in their analyses (Goldfuss 2016). By removing this requirement, the Trump administration eased the approval of federal projects that were known to negatively impact climate change.

Withdrawal from the Paris Agreement

On June 1, 2017, Trump announced U.S. withdrawal from the Paris Agreement (UVA/Miller Center 2022). The Paris Agreement was the then current international treaty implementing the Framework Convention on Climate Change. Under the Obama administration, the

U.S. had submitted its Nationally Determined Contribution (NDC) of GHG reductions from 26–28 percent below 2005 levels by 2025. Trump stated that "as of today" the U.S. would cease all implementation of the Paris Agreement. Trump's withdrawal was a bit of a technicality. The Paris Agreement does not allow parties to submit their notice of withdrawal until three years after the treaty comes into force, on November 4, 2016. After that, there is a 12-month period before withdrawal becomes effective, so the earliest the U.S. could complete withdrawal would be November 4, 2020 (Sabin Center for Climate Change Law 2022). Despite losing the re-election bid in 2020, Trump's earlier withdrawal efforts became a fait accompli, making the U.S. the only country to withdraw.

The U.S. pull out was received with disgruntlement both at home and abroad. In 2020, the U.S. contributed 15 percent of the world's GHG emissions. Under the Trump administration, U.S. GHG emissions initially rose but given efforts of the states and localities, decreasing prices of natural gas displacing coal, and the economic downturn during the pandemic, emissions dropped to slightly less than they were when Trump took power. Still, the decision not to attempt to reduce emissions had consequences in terms of time lost in promoting faster reductions. Withdrawal also gave the U.S. an image of being an unreliable partner, although for the Trump administration that criticism was applied more broadly than to just environmental issues (BBC News 2020).

While the U.S. announced its withdrawal from the Paris Agreement, U.S. delegates continued to attend the annual conference of parties that were held after. In Bonn, at COP23, the State Department delegation was smaller and there was little engagement in negotiations (Finnegan and Ebbs 2017). The U.S. would not again become an eager participant in the annual conferences until Biden became president-elect in November of 2020.

Revoking the Federal Sustainability Plan

On May 17, 2018, President Trump issued an executive order revoking an Obama EO 13693 which had established the Federal Sustainability Plan. Under that plan, each government agency was to adhere to the goal of reducing GHG emissions by 40 percent over a decade. Under the new executive order, agencies were not required to consider GHG emissions in any way (Sabin Center for Climate Change Law 2022). This effort was widely seen as an attempt to boost the fossil fuel industry, especially coal interests, the largest contributors to GHG emissions.

The California Waiver

In September of 2019 the Trump administration revoked California's authority to set its own tailpipe standards and to mandate electric vehicles. California's unique right to set stricter air quality standards dates to the 1970s with the passage of the Clean Air Act (Popovich and Plumer, What Trump's Environmental Rollbacks Mean for Global Warming 2020). Under that law, California was permitted to seek a waiver of the preemption which prohibits states from issuing emission standards for new cars. The Clean Air Act allowed other states to adopt California's emission standards without seeking separate EPA approval (EPA 2022). California's standards were developed before the passage of the Clean Air Act to address its own unique air quality issues. California's standards are stricter than federal standards, thus driving the state, and other states that adopt California's standards, to cleaner air. Trump's denial of the waiver meant dirtier air and less pressure on automakers to produce clean vehicles, thus contributing to climate change.

Reversal on Obama-Era Fuel Efficiency Standards

In March of 2020, the Trump administration, working through the EPA and the Department of Transportation (DOT) finalized a rule that rolled back tailpipe emissions. The new rule allowed cars to emit nearly a billion tons of CO^2 over their lifetime, more than the Obama rule would have allowed. By allowing more emissions, the Trump administration did much to harm climate policy and consumers as well, considering that under the new rule consumers would end up paying more for gasoline. Under the new rule, instead of fleets having to average 54 miles per gallon by 2025, fleets would only have to average 40 miles per gallon. Automakers generally did not support the change, believing it would result in years of uncertainty as the new rule was contested in the courts (Davenport 2020a).

Methane Leaks

The Trump administration revised two major Obama-era rules on methane emissions from oil and gas operations. Methane, the main ingredient in natural gas, contributes 36 times more potently to climate warming than CO^2. This rollback was important as scientific studies showed that methane more commonly leaks from oil and gas operations than was previously thought. In addition, the Trump administration also delayed rules on methane emissions from landfills, another substantial source of emissions (Popovich and Plumer, What Trump's Environmental Rollbacks Mean for Global Warming 2020).

Cost-Benefit Rule

In a flurry to finalize more anti-environmental rules before leaving office, in December of 2020, the Trump administration finalized a rule that changed the way cost-benefit analyses are used in economic analyses of clean air and climate regulations. The rule was intended to give industry more flexibility in challenging EPA regulations in the courts; however, coming so late in the administration the anticipation was that the Biden administration would quickly reverse the rule (Davenport 2020b).

Established a Minimum Pollution Threshold for GHG Emissions

Also, in a last-minute change when leaving the administration, the EPA finalized a rule that established a minimum threshold at which EPA can regulate GHG emissions from stationary sources. That level was set at 3 percent of total U.S. GHG emissions. This threshold included power plants but let oil and gas facilities off the hook (Popovich, Albeck-Ripka and Pierre-Louis, The Trump Administration Rolled Back More Than 100 Environmental Rules. Here's the Full List 2021).

Climate Super Pollutants

On February 26, 2020, EPA finalized a rule that relaxed standards for refrigeration and cooling companies must follow in checking for leaks of hydrofluorocarbons (HFCs), a power GHG. The new rule rescinded a 2016 regulation that extended regulation of HFCs. Under the new rule, appliances with 50 or more pounds of alternative refrigerants like HFCs would no longer be required to undergo leak inspection (Sabin Center for Climate Change Law 2022). Legislative action, discussed below, would change the dynamics on HFCs and bring them back into a better scenario for climate change policy action.

Arctic National Wildlife Refuge (ANWR)

In one of its last acts before leaving office, the Trump administration opened ANWR to oil and gas leasing. On January 6, 2021, the Bureau of Land Management opened bidding on the coastal plain of

ANWR. BLM received 13 bids covering more than 550,000 acres, some in the sensitive 1002 Area, and bringing in more than $14 million in fees (BLM 2021). The ANWR had been fought over for decades, with environmentalists arguing that opening the area to drilling would threaten the highly sensitive animal populations. However, oil interests had long sought to drill there because reserves were thought to be great. In 2015, the Obama administration had released a 15-year management plan for the area that included keeping it wilderness. The Trump administration, though, had a provision written into the Tax Cuts and Jobs Act of 2017 that required the government to hold at least two oil and gas lease sales before 2027 (Rahm 2019).

Trump and the Courts

There are two aspects of the Trump administration's interaction with the courts that should be considered. The first is litigation over the way Trump-era agencies used the agencies' powers to implement Trump's policies. During the Trump administration there was considerable litigation over agency rules and guidance. The second was the attempt on the part of the Trump administration and the Republicans to pack the courts with ideologues who shared the Republican policy agenda of deregulation and limiting the power of federal agencies.

Trump-Era Litigation

Over the course of the Trump administration and in the months after as the cases moved through the courts, the overwhelming number of court outcomes were against the Trump administration's actions. In the area of climate change policy, the major outcomes included several of significance. For instance, in 2021 (as mentioned earlier) the courts vacated the Affordable Clean Energy Rule, Trump's replacement of Obama's Clean Power Plan. This issue was left to the Biden administration to settle. In 2020, the courts vacated the Department of the Interior office of Ocean Energy Management's approval of an offshore oil drilling facility off the coast of Alaska. In 2020 the courts also blocked numerous oil and gas leases on federal lands in Wyoming, telling the agency it needed to reconsider the GHG emissions as required by the National Environmental Protection Act (Institute for Policy Integrity 2022).

A major loss for the Trump administration was the rejection by the courts of the Department of the Interior's attempt to rescind protections against methane waste on public and tribal lands. In July of 2020, the courts ruled that the Bureau of Land Management's justification for the 2018 "recission rule" was inadequate. The court ruling was a major win for the states of California and New Mexico which had fought in the courts since the rule was adopted for its reversal (Farah 2020).

Those in opposition to the Trump administrations deregulatory agenda filed hundreds of lawsuits across the country during the Trump years. In 378 of the cases where climate issues were at the center, nearly 90 percent sought to stop deregulatory efforts and to strengthen environmental protections. Just about 10 percent sought to strengthen deregulatory efforts. Non-governmental organizations (NGOs) filed the most cases, followed by government entities, industry, individuals or labor groups. The federal government was the defendant in nearly 90 percent of these cases although states and localities were also targets. Of the cases that sued industry, the vast percentage of defendants were fossil fuel companies (Silverman-Roati 2021).

Packing the Courts

Trump showed his intentions in regard to the U.S. Supreme Court with the 2017 appointment of Neil M. Gorsuch, sworn in as Justine Antonin Scalia's replacement (UVA/Miller Center 2022). Gorsuch was the son of Anne Gorsuch, the EPA Administrator in the Reagan administration, whose leadership at the EPA was driven by her attempts to downsize the agency and reduce its regulatory reach. Her opposition to spending Superfund monies appropriated by Congress eventually led to her firing by President Reagan in an attempt to calm the outrage she provoked.

Neil Gorsuch followed in his mother's steps. While an appeals court judge, Neil Gorsuch had written that the Chevron deference, which refers to the prior position of the Supreme Court to defer to the expertise of bureaucrats, allows bureaucrats to "swallow huge amounts of core judicial and legislative power" (Cohen 2022). By appointing Gorsuch to the U.S. Supreme Court, the Trump administration had secured one large vote for its agenda of deregulation and reduction of the power of administrative agencies such as EPA.

Trump's second opportunity to shape the U.S. Supreme Court came with the nomination of Brett Kavanaugh after Justice Anthony Kennedy announced his resignation (UVA/Miller Center 2022). Kavanaugh had a track record of anti-environmental and anti-public health rulings. In his 12 years on the United States Court of Appeals for the District of Columbia, Judge Kavanaugh voted to limit the authority of EPA rules on climate change and air pollution. His legal philosophy hinged on the idea that lacking explicit instructions from Congress, any far-reaching attempts to limit pollution should be met with extreme caution by the courts. Trump's appointment of Kavanaugh, replacing the prior swing vote of Kennedy, moved the court to the right (Plumer 2018).

On September 18, 2020, U.S. Supreme Court Justice Ruth Bader Ginsburg died. Despite the opening's being only seven weeks from election day, the Republicans pushed to nominate a replacement. By the 26th, Trump announced his Supreme Court nomination: Judge Amy Coney Barrett, a conservative Federal Appeals Court judge. On October 26, 2020, the Senate voted to confirm Barrett, just eight days before the election (UVA/Miller Center 2022). Barrett's views on environmental issues were not widely known before she was seated on the court, but her prior clerkship with Justice Scalia suggested she would become the sixth conservative vote on such issues (Grandoni 2020).

While the results of the Supreme Court appointments would not shape policy until their first-term case results were released, it was clear at the end of the Trump administration that conservative control over the judiciary had been achieved. This was not just in the Supreme Court, but across the board in the federal court system. Some of this had been achieved even before Trump took office but with Trump in the presidency a large number of conservative judges were appointed. In 2018, Senate Republican Leader Mitch McConnell (R-KY) made his strategy clear in a speech given before the Federalist Society. He said his aim was to do everything he could to transform the judiciary because he argued, everything else is transitory.

McConnell's refusal to hold hearings on Obama's Supreme Court nominee, Merrick Garland, was just one example of his strategy. Over the years, the Republicans in the Senate blocked consideration of any Democratic nomination to the federal courts, leaving a huge vacancy of positions in the federal district and appeals courts. Trump was able to successfully appoint, and get confirmed, 226 judges. All these judges shared a common ideological approach: ending the wall between church and state, reducing the power of the federal agencies, banning abortion, and expanding gun rights (Calmes 2021).

Intergovernmental Panel on Climate Change (IPCC) Reports During the Trump Administration

The IPCC released several reports during the Trump administration that continued the pattern of increasing scientific certainty and concern over climate shifts. The first of these was a special report released in 2018 entitled *Global Warming of 1.5° C* which emphasized that the globe had already experienced 1.0° C increases in temperature since pre-industrial levels and that temperatures would likely reach 1.5° C between 2030 and 2052. The IPCC stressed that the world had only twelve years left to limit emissions to keep temperature rise below 1.5° C. The report stressed that global impacts of an increase of 1.5° C would be far less than if the world hit the 2.0° C threshold (IPCC 2018). The urgency of these warnings was bold and strongly stated.

The IPCC also released updates on the methodologies to assess GHG emissions, and special reports on climate change and land, and the ocean and the cryosphere in a changing climate (IPCC, Reports 2022). But the most important report released during the Trump era was the first volume of the Sixth Assessment Report (AR6) called *AR6 Climate Change 2021: The Physical Science Basis*. The report included many direct statements for policymakers including: "It is unequivocal that human influence has warmed the atmosphere, ocean and land. Widespread and rapid changes in the atmosphere, ocean, cryosphere and biosphere have occurred" and "the scale of recent changes across the climate system as a whole—and the present state of many aspects of the climate system—are unprecedented over many centuries to many thousands of years" (IPCC, Summary for Policy Makers 2021).

In addition, the report stated that anthropogenic-induced climate change was already affecting many weather and climate extremes in all regions of the world and that these extremes include heatwaves, heavy precipitation, hurricanes and cyclones, and droughts. The report made clear that scientific certainty of attributing these extremes to human influence had increased since the release of AR5. Finally, the report alleged that "Low-likelihood outcomes, such as ice-sheet collapse, abrupt ocean circulation changes, some compound extreme events, and warming substantially larger than the assessed very likely range of future warming, cannot be ruled out..." (IPCC, Summary for Policy Makers 2021).

The Opposition
During the Trump Administration

The interests who opposed climate change action had a champion in Donald Trump at the federal level. Republicans also had command of many state legislatures and governorships leading to policies of deregulation and the promotion of fossil fuels which tended to undermine environmental protections and climate policies. The fossil fuel sector, at least portions of it, gained a great deal during the years and policies of the Trump administration. The rollback of clean car fuel efficiency standards, alone, profited the oil and gas industry by as much as $200 billion. The sector continued to pour vast sums into lobbying in favor of Trump's call for "energy dominance." Top officials in the oil and gas sector contributed more than $56 million to the 2016 election and they worked, successfully, to place top executives of the oil and gas sector into positions within the administration (Hardin 2018). However, despite Trump's rhetoric to promote "beautiful clean coal," coal jobs at the end of the Trump administration were down by 24 percent (Kuykendall 2020). This was due to market forces. The abundance of cheaper natural gas drove a wave of coal plant closures.

Pushback During the Trump Administration

There was some pushback on the advance of the opposition. For instance, in November of 2018, the Democrats took control of the House of Representatives, with a balance of power of 225 to 197. This provided a forum for detractors of the Trump administration. The first opportunity arose in September of 2019 when a whistleblower complaint, filed by a member of the U.S. intelligence community, was made public. It alleged that the president was trying to get foreign interference to help him win the 2020 election. After this revelation, Speaker of the House Nancy Pelosi announced that the House would begin a formal impeachment inquiry against President Trump. The specific issue was a phone call between Trump and Ukrainian President Zelensky in July in which Trump pressured the Ukrainian president to provide Trump with information that could help discredit a domestic political rival, Joe Biden. The House of Representatives impeached President Trump in December of 2019, but because the Republicans held the majority in the Senate, no conviction was forthcoming.

Trump became the only U.S. president to be impeached twice. The second came after the January 6, 2021, mob attack on the U.S. Capitol. The House impeachment was for incitement of insurrection. Again, Trump was acquitted by the Senate (UVA/Miller Center 2022). Because the House majority was Democratic, this enabled more progressive members of the Democrats to introduce legislation they wanted to see considered. Representative Alexandria Ocasio-Cortez took advantage of this and in 2019 introduced H. Resolution 109 calling for the Green New Deal (discussed later).

As society became more divided during the Trump administration, cleavages could be clearly seen on many issues, including climate policy. Along with the hundreds of legal actions taken to challenge Trump's environmental rollback of Obama-era environmental protections, there was the widespread growth of the youth climate movement. The movement began internationally in 2018, when Greta Thunberg and other activists protested before the Swedish parliament to demand climate action. This movement led to the organization of Fridays for Future, a youth-led and -organized movement for climate action (Fridays for Future 2022).

Within the U.S., the Sunrise movement stared in 2017, with the initial goal of making climate change an issue of importance in the 2018 midterm elections (Matthews and Hulac 2018). Sunrise spread throughout the country with the creation of more than 400 hubs linked together through information and communication technology. Sunrise is youthled and -run. It is decidedly political in its orientation. It opposes the Republicans and their alliance with the fossil fuel industry. It endorses candidates for political office. Its key issue is climate change, which it wants to fight by creating millions of good green energy jobs and a good future for all Americans (Sunrise Movement 2022).

The Coronavirus Pandemic

Beginning in 2020, the pandemic considerably disrupted national affairs. President Trump declared a national emergency on March 13, 2020. The United States led the world with cases. As states and localities tried to limit activities to reduce the spread of the disease, the economy was widely disrupted in the largest downturn since the Great Depression. Unemployment in the U.S. hit 22 million (UVA/Miller Center 2022). But the pandemic also presented some opportunities for passage of legislation previously considered unthinkable.

In the waning days of the Trump administration, Congress authorized a $900 billion coronavirus relief package that included $35 billion in spending on wind, solar, batteries, and carbon dioxide reduction technologies. But, most importantly, the provision cut the use of hydrofluorocarbons refrigerants, powerful emitters of GHGs. The law required manufacturers to cut production of HFCs by 85 percent by 2015. This law became the first significant climate change law passed since 2009. The relief package was backed by a several powerful Congressional Republicans, making it a pushback on the four years of climate denial by the Trump administration. The refrigerant provisions were supported by the chemical industry in large part because they had developed alternatives to HFCs and a phaseout would give them a competitive advantage over manufacturers using the older technologies.

Additionally, the chemical industry had grown concerned that states might act to phase out HFCs independently in a patchwork of rules that would make it harder for the industry than one national standard. The push by industry gained the support of Republicans for the HFC provisions (Davenport 2020c). The bipartisan passage of the HFC provisions signaled that the Senate would likely ratify the Kigali Amendment to the Montreal Protocol, placing it among some of the earliest measures considered under the Biden administration.

Regional, State, and Local Action

Many regional climate chance alliances continued under the Trump administration. The Regional Greenhouse Gas Initiative (RGGI) continued operating in the northeastern U.S. In 2017 it adjusted its cap for its mandatory cap-and-trade program for the electricity sector downward by 30 percent from 2020 levels by 2030. New Jersey had left RGGI in 2012 under the leadership of Republican governor Chris Christie but returned in 2019 when the newly elected Democratic governor, Phil Murphy, took power. The Western Climate Initiative (WCI) continued with California and Quebec as members.

The U.S. Climate Alliance was newly formed in 2017 by the governors of California, New York, and Washington shortly after President Trump announced U.S. withdrawal from the Paris Agreement. Other states joined over time and by the end of the Trump administration 25 states were members. When agreeing to join, states declared their commitment to reaching the Paris Agreement target of keeping temperatures rises below 1.5° C. To do this, states pledged to reduce collective net GHG emissions by at least 26–28 percent by 2025 and by 50–52 percent by 2030 below 2005 levels and to collectively achieve and overall net-zero GHG emissions no later than 2050. Other regional groups included the Governors' Accord for a New Energy Future which formed as the Trump administration was coming to power. The agreement pledged states to commit to a clean energy future and was joined by California, Connecticut, Delaware, Hawaii, Iowa, Massachusetts, Michigan, Minnesota, Nevada, New Hampshire,

New York, Oregon, Pennsylvania, Rhode Island, Vermont, Virginia, and Washington.

The Transportation and Climate Initiative (TCI) was originally launched in 2010 by 12 jurisdictions to mitigate the transportation sectors' impact on climate change (among other goals). In 2019, the TCI states included Connecticut, Delaware, Maine, Maryland, Massachusetts, New Hampshire, New Jersey, New York, Pennsylvania, Rhode Island, and Vermont, plus the District of Columbia. In 2019 the TCI members released a draft Memorandum of Understanding (MOU) which they finalized in 2020. The MOU provided for implementation by 2022 (Center for Climate and Energy Solutions 2022).

Various states were also individually active. For instance, California reached a deal with automakers (Ford, Honda, BMW, Volkswagen, and Volvo) to maintain Obama-era fuel economy standards of 54 miles per gallon, despite Trump's rollback of the rule. Colorado and New Mexico created policies to curb methane leaks (Popovich and Plumer, What Trump's Environmental Rollbacks Mean for Global Warming 2020). California also passed SB100 in 2018 which required the power grid to be carbon free by 2045. A number of states created GHG emission standards for the first time, especially when the Democrats gained a majority in the state legislature. Two prominent ones were Colorado and New York. Another area of movement during the Trump years was for states to upgrade or put in place Renewable Portfolio Standards (RPS). Maryland, for instance, upgraded its RPS in 2019 requiring that 50 percent of the portfolio mix be renewable by 2030. New York went even further, requiring that 70 percent of its fuel mix be renewable by 2030 (Bromley-Trujillo and Holman 2020).

Not all state action resulted in favorable climate change policies. For instance, in 2018 Washington failed to pass a ballot initiative which would have created a carbon tax. It was largely defeated because the electric sector fought the measure strongly. Oregon's attempt to pass a cap-and-trade measure also failed when minority Republican members of the state legislature took advantage of a little-used super-majority quorum requirement and walked out in 2019 and 2020, thus preventing action on climate legislation that would otherwise have passed. Given the strength of the utility industry in Ohio, the state passed HB-6 in 2019 which subsidized nuclear and coal power plants through a surcharge on consumers. Ohio also cut its RPS and its energy efficiency programs. In 2019, Kentucky cut the rate at which a solar rooftop generator of electricity could sell it to the grid via net metering programs. Indiana passed legislation that would fully phase out net metering by 2022 (Bromley-Trujillo and Holman 2020).

Conclusion

After four years of the Trump administration, federal environmental policy had been compromised and climate policy was abandoned. The Trump administration had done its best to unravel all of Obama's accomplishments, although those done through administrative action alone were vulnerable to reversal in the next administration.

The Trump administration left office following a chaotic series of events. On Election Day, November 3, 2020, Trump carried 23 states and 213 Electoral College votes. The race was close in many states and no winner was declared on election night in large part because of the huge number of mail-in votes. Trump, however, declared victory and claimed that the remaining ballots not yet counted should be tossed out due to fraud. On November 7, 2020, the press called the election for Joe Biden. Trump refused to accept the result and his campaign team filed lawsuits across swing states in an attempt to contest the results due to voter fraud from mail-in ballots.

On December 13, 2020, the U.S. Supreme Court rejected the case brought by Texas and 17 other states alleging widespread voter fraud. The next day, William Barr resigned as Attorney General after Trump discussed firing him over the failure of the Justice Department to find fraud in the election results.

On December 16, 2020, the Electoral College declared Joe Biden the 46th president. But Trump's efforts continued. On December 20, 2020, the Justice Department under U.S. Attorney General William Barr announced it had done a detailed investigation and had found no widespread voter fraud. On January 2, 2021, Trump called the Secretary of State of Georgia insisting that he "find" the votes necessary for Trump to win Georgia. Tensions mounted. On January 6, 2021, Trump supporters gathered at the Capitol and a mob overwhelmed the Capitol police. On January 13, 2021, the House of Representatives impeached Trump for a second time—this time for "incitement of insurrection." Trump left the White House on the morning of January 20, 2021, without attending the inauguration of Joe Biden. Trump thus became the first president to fail to oversee the peaceful succession of power in the U.S. since 1869 (UVA/Miller Center 2022).

CHAPTER 7

The Pendulum Swings Again

The Joseph R. Biden Administration

Introduction

Joseph R. Biden was sworn in as the 46th president on January 20, 2021. All attending wore masks due to the ongoing Covid-19 pandemic. The scene was also dramatized by the presence of almost 25,000 national guard troops deployed out of fear of additional violence. It had been only two weeks since the mob assault on the Capitol. Out-going Vice President Pence was in attendance as were the Clintons, the Bushes, and the Obamas. Donald Trump did not attend. He had left the White House that morning, on his way to Florida. Work for the Biden administration began just hours after his inauguration, with the new president signing 17 executive orders aimed at reversing many of the most controversial positions of the Trump administration. The Biden administration began with Democrats in control of the Senate, albeit only with the deciding vote from Vice President Harris. Democrats also controlled the House (Gambino 2021).

The Biden administration came to office under the full weight of the pandemic. When Biden took office, unemployment was 6.2 percent. Although high, it was down from the 14.7 percent rate in March of 2020 (Bureau of Labor Statistics 2022). The first Consolidated Appropriations Act (CARES) had been signed into law December 27, 2020, just a month before Biden took over the presidency. It was a series of programs related to the pandemic that provided fast and direct financial assistance to families and workers. Despite its positive impact, a second CARES Act was signed into law in March of 2022 again to address the huge financial impacts of the pandemic (U.S. Department of Treasury 2020). In March of 2020 the states had begun shutdowns to slow the pace of the pandemic. As the pandemic worsened in the spring of 2020, shortages of personal protective equipment (PPE) began to appear. Many hospitals were overwhelmed with the number of infected. In November of 2020, just after the election of Joe Biden, cases spiked in the U.S.

In December of 2020, the first vaccine against Covid-19 was authorized by the CDC. Shortly after Biden took power, the virus had infected more than 100 million worldwide and U.S. deaths had reached half a million (CDC 2022). The pandemic's toll would continue to weigh on the Biden administration well into May of 2022 when the U.S. had officially registered more than one million deaths (Donovan 2022). In the U.S., by summer of 2022 the Omicron BA.5 wave of the pandemic finally started to ebb (Flam 2022). By then, inflation had begun to impact the U.S. economy, reaching a high of 9.1 percent by June of 2022, a forty year high (BLS 2022). In the midst of this horrible situation, the Biden administration took office and began to frame policy shifts. Environmental policy and climate policy, in particular, were key foci.

Governmental Appointments

President Biden appointed a team of cabinet secretaries and administrators that shared his view on reversing Trump's climate and energy positions. Central to these appointments was that of former Secretary of State John Kerry to serve in a newly created cabinet position of Special Climate Envoy. In that role, Kerry was supposed to reach out to foreign nations to signal that the U.S. was resuming its role in climate leadership. This appointment was announced before the inauguration, showing the new administration's emphasis on reversing the Trump climate legacy. By making Kerry a member of the National Security Council, the Biden administration indicated that it viewed climate issues at the high level of national security (Friedman, With John Kerry Pick, Biden Selects a "Climate Envoy" With Stature 2020).

In a similar manner, Biden named Gina McCarthy, the former Administrator of the Environmental Protection Agency (EPA), to be National Climate Advisor. In that role, McCarthy was charged with coordinating efforts across the federal government to quickly lower greenhouse gas (GHG) emissions. This appointment flowed from a commitment Biden made while campaigning, to reduce U.S. GHG emissions and to be carbon neutral by 2050 (Detrow, Keith and Ludden 2020). Gina McCarthy held this post until the Biden administration was successful in passing the Inflation Reduction Act of 2022 (IRA) (discussed below), which contained over $370 billion in funding for clean energy. She was succeeded by her deputy, Ali Zaidi. As part of the reshuffle, John Podesta was brought in to oversee the federal spending on clean energy under the law. Zaidi was also to serve as vice chairman of a task force of cabinet secretaries led by Podesta (Friedman, Biden, Remaking Climate Team, Picks John Podesta to Guide Spending 2022).

Other cabinet positions also signaled a new direction. These included the appointment of Pete Buttigieg to run the Department of Transportation. As a candidate for the Democratic presidential nomination in 2020, Buttigieg drafted a climate plan with the goal of getting to net zero carbon emissions by 2050 (Committee for a Responsible Federal Budget 2020). The U.S. transportation sector surpassed electric generation in GHG emissions in 2017, making the role of heading that agency critical to any climate change plan.

Biden appointed Jennifer Granholm to run the Department of Energy (DOE). Before becoming the Secretary of DOE, she was governor of Michigan. In her role as governor, she established the 35-member Michigan Climate Action Council and charged it with developing a plan for Michigan that would not only address climate change but also do it in a way that would create jobs and grow the economy. The Michigan plan was tightly focused on building jobs in the renewable energy sector and implementing expansion of renewable energy use in the state (Michigan.gov 2007). She brought these views to the job of Secretary of DOE. She also brought her experience as a two-term governor of a state dominated by the auto industry—an industry central to the energy transformation necessary to confront climate change.

Biden appointed Michael Regan as EPA Administrator. Regan became the second person of color to run the agency. Prior to his appointment as EPA chief, Regan served as the Secretary of the North Carolina Department of Environmental Quality (DEQ). While at DEQ, Regan developed that state's climate and clean energy plan. Regan also had prior experience as the Southern Regional Director of the Environmental Defense Fund. In that role, he worked to partner with communities and industries to create pragmatic solutions to climate change. Regan had begun his career with the EPA and eventually became a national program manager partnering with industry to reduce air pollution, improve energy efficiency, and address climate change (EPA, Michael S. Regan 2021).

Biden's choice to head the Department of the Interior (DOI) was Deb Haaland, the first Native American to hold such a post. Before serving as Secretary of Interior, Haaland was a member of Congress who focused on environmental justice and climate change in addition to issues of importance to Native Americans (DOI, Secretary Deb Haaland 2021).

Another key member of the Biden team included Secretary of

State, Antony Blinken. Blinken represented the U.S. on the UN Security Council's meetings on Climate and Security as well as representing the U.S. on all bilateral and multilateral diplomatic discussions where he was charged by the president to include consideration of U.S. climate goals in each negotiation (Blinken 2021).

All in all, the team that Biden assembled shared a concern for the environment and a particular concern over climate change. They would articulate a pro-climate agenda across the federal agencies and present that image in international dialogue with the goal of regaining the leadership position of the U.S. that Trump had relinquished.

Reversing Trump's Agenda Through Executive Action

Domestic Executive Actions

On his first day in office, Biden issued a series of executive orders that marked the shift in policy. First, Biden rejoined the Paris Agreement. By rejoining the agreement, Biden was announcing to the world not only that the U.S. was going to abandon "America First" and follow the international rules but also that the U.S. was returning to a commitment to multinationalism. By returning to the agreement, Biden committed the U.S. to improve the Nationally Determined Contribution (NDC) first presented in 2015 (McGraath 2021).

Also, Biden revoked the Keystone XL pipeline permit that Trump had brought back after Obama's denial of it. Biden stated that building the pipeline would not be consistent with his climate agenda. Canadian Prime Minister Justin Trudeau responded by saying that the fulfilling of President Biden's campaign promise was a disappointment; however, Trudeau praised the return of the U.S. to the Paris Agreement (Gillies 2021).

The new president declared a 60-day suspension of new oil and gas drilling on federal lands, further indicating a shift away from fossil fuels (Groom and Hiller 2021). On his first day in office, Biden also re-established the Social Cost of Carbon Interagency Task Force by executive order (Rennert et al. 2021). Biden issued a comprehensive executive order requiring all federal departments and agencies to review all actions taken during the Trump administration that conflicted with Biden's national policy goals and to take steps to reverse them.

The national policy goals elaborated by the Biden administration included "to listen to the science; to improve public health and protect our environment; to ensure access to clean air and water; to limit exposure to dangerous chemicals and pesticides; to hold polluters accountable, including those who disproportionately harm communities of color and low-income communities; to reduce greenhouse gas emissions; to bolster resilience to the impacts of climate change; to restore and expand our national treasures and monuments; and to prioritize both environmental justice and the creation of the well-paying union jobs necessary to deliver on these goals" (Biden, Executive Order on Protecting Health and the Environment and Restoring Science to Tackle the Climate Crisis 2021).

On January 27, 2021, Biden issued an executive order named "Tackling the Climate Crisis at Home and Abroad." In this executive order the president stated that the climate crisis was at the center of U.S. foreign policy and national security. To regain U.S. leadership, he said he would call a summit meeting of world leaders ahead of the Glasgow COP26 meeting to raise awareness of the climate crisis and create ambitions for the Glasgow meeting. He ordered a government-wide approach to the climate crisis. Biden created The White House Office of Domestic Climate Policy in the Executive Office of the President with the job of coordinating domestic climate policy.

He created a National Climate Task Force with members from the cabinet agencies as well as EPA, the Council on Environmental Quality (CEQ), the Office of Science and Technology Policy (OSTP), and Office of Management and Budget (OMB), with the mission of implementing a government-wide approach to the climate crisis. He promised to use the government's purchasing power to promote his agenda including establishing a clean electricity and vehicle purchasing program.

He instructed the Secretary of the Interior to review how more clean energy could be sited on federal lands and waters. He permanently paused oil and gas leasing on public lands until a comprehensive review could be undertaken. He told agency heads to review their expenditures on gas and oil subsidies and to eliminate them in their Fiscal Year 2022 budget requests and thereafter. He ordered agencies to submit plans to improve climate resilience in their facilities and buildings. He promised to create a Civilian Climate Corps to address many issues of clean air and water. Finally, he created the White House Environmental Justice Interagency Council and tasked it with developing a strategy to address current and historic environmental injustice (Biden, Executive Order on Tackling the Climate Crisis at Home and Abroad 2021). Later in the day, a press conference was held in the White House by John Kerry, the Special Climate Envoy, and Gina McCarthy, the National Climate Advisor, to publicize the contents of the executive order. They said the order issued that day built on the actions taken on the first day of the administration to fight the climate crisis. They elaborated on the issues in the executive order and took questions from the press concerning the administration's agenda on climate change (The White House 2021).

The Biden administration made an enormous commitment to environmental justice when Biden issued executive order 14008, called the Justice40 Initiative. It established the goal that 40 percent of overall benefits of certain federal investment flows would be aimed at minority and disadvantaged communities. The kinds of investments that fell under its jurisdiction included climate, clean energy, energy efficiency, clean transit, sustainable housing, and remediation and reduction of historic pollution (The White House 2021).

On March 9, 2022, EPA rescinded the decision promulgated by the Trump administration that denied California its waiver from the Clean Air Act. California is allowed, under the Clean Air Act, to seek a waiver of preemption which prohibits the states from enacting emission standards for new vehicles (EPA, Vehicle Emissions California Waivers and Authorizations 2022). On March 14, 2022, EPA issued a notice of the decision to restore California's waiver that had been denied by the Trump administration. This action gave California, and other states that wished to adopt California standards, the right to set and enforce more stringent emission standards and its zero emissions mandate (Cattaneo 2022).

In April 2020, the Biden administration finalized rules to toughen fuel efficiency standards that had been rolled back during the Trump administration. Under the new rules, the National Highway Traffic Safety Administration (NHTSA) increased fuel efficiency standards by 8 percent in both 2024 and 2025 model years and 10 percent in 2026 model year. The EPA's stricter vehicle emissions targets, finalized in December, parallel NHTSA's guidelines. The EPA rules result in an average of 40 miles per gallon, as opposed to Trump's 32 miles per gallon (Reuters 2022).

On June 1, 2021, the Biden administration suspended oil and gas drilling in the Arctic National Wildlife Refuge (ANWR). Biden's actions on his first day in office had stopped new drilling and this decision paused leases granted by the Trump administration pending a review of their environmental impacts and the legality of the Trump administration's decision to grant them. While the decision was met with celebration from environmentalists, it was opposed by Alaskan elected officials. Senator Lisa Murkowski (R-AK) argued that the leases had been part of the Tax Cut and Jobs Act of 2017 and that the administration did not have the authority to suspend them, even temporarily

(Davenport, Fountain and Friedman 2021). This action on the part of the administration did not permanently end drilling in ANWR but it slowed it by creating obstacles that the companies holding leases and the pro-drilling politicians would have to fight.

In August of 2022, a ban on coal mining on federal land was reimposed by the courts. The temporary ban reversed the Trump administration's resumption of coal mining on public land that has been stopped during the Obama administration (Phillips 2022). The ruling left it to the Biden administration to review the action and determine whether coal leasing should resume.

On September 22, 2022, the EPA finalized a rule to phase out the use of hydrofluorocarbons (HFCs). The rule was drafted as a result of the American Innovation and Manufacturing (AIM) Act of 2020, which was passed on December 27 of 2020 when Biden was president-elect. The law provided new authority for the EPA to limit HFCs through an allowance and trading program. The law required EPA to phase out HFCs by 85 percent by 2036, beginning in 2022 with a 10 percent reduction (EPA, Phasedown of Hydrofluorocarbons Final Rule Frequently Asked Questions 2022).

The EPA action was followed in September of 2022 with Senate ratification of the Kigali Amendment to the Montreal Protocol of 1987, which requires countries to put in place HFC controls similar to what the U.S. adopted in the AIM Act. Passage of the Amendment was bipartisan and strongly backed by the manufacturing sector, which expected to profit enormously in exports of replacement chemicals. The Amendment was also heavily supported by the environmental advocates, who wished to see the production and use of HFCs drastically reduced due to the fact that they are hundreds to thousands times more powerful than carbon dioxide (CO^2) in warming the planet (Mufson, U.S. Ratifies Global Treaty Curbing Climate Super-Pollutants 2022).

In October of 2022, the Biden administration announced the expansion of offshore wind energy. The Department of the Interior's (DOI) Bureau of Ocean Energy Management (BOEM) announced plans for sale of leases off the coast of California, the first leases on the West Coast. This opened the potential for commercial-scale offshore floating wind energy development (DOI, Biden-Harris Administration Announces First-Ever Offshore Wind Lease Sale in the Pacific 2022).

In November of 2022 the EPA announced a rule tightening control over methane leaks coming from oil and gas operations. The new proposed rule would update and strengthen a prior rule from 2021 (EPA 2022). Scientists consider this update of great importance because methane is one of the most powerful GHGs being emitted. Control over methane emissions would be a very important factor in reducing warming.

International Engagement

On March 26, 2021, President Biden invited 40 world leaders to a virtual climate summit to be help on April 22–23, 2021. Invitees included Chinese President Xi Jinping and Russian President Vladimir Putin (The White House 2021). The meeting took place as planned under the name the 2021 Leaders Summit on Climate. It was attended by the world's 17 largest economies and was undertaken to underscore the importance of the Glasgow COP26 to be held in November (U.S. Department of State 2021).

President Biden attended the Climate Summit in Glasgow, Scotland, in November of 2021. The meeting was COP26 of the United Nations Framework Convention on Climate Change (UNFCCC). While there, Biden apologized for Trump's exit from the Paris Agreement. The large contingency of U.S. delegates, which included John Kerry and Gina McCarthy, signaled the importance the U.S. saw in the meeting (Adam, Dennis and Linskey 2021). Before leaving the leaders conference, which took place on the first two days of the meeting, Biden spoke sharply against both China's Xi Jinping and Russia's Vladimir Putin for not attending, saying that they had abandoned any potential leadership position on climate change.

China, the world's biggest emitter of GHGs, had presented a plan on the eve of the conference that disappointed many climate analysts. The plan did not deviate from an earlier version China had submitted to the world and called only for emissions to peak by 2030 before falling to net-zero by 2060. However, if such a schedule was followed, the goal of achieving only a 1.5° C temperature increase would be put in jeopardy (Harvey 2021).

Biden also attended COP27, held in Sharm el-Sheikh, Egypt in November of 2022. The UN Secretary-General called the meeting to order with a stark warning. António Guterres set the tone by saying the world was "on a highway to climate hell with our foot on the accelerator." He added that "we are in a fight for our lives, and we are losing." The talks opened under the shadow of new data released by the World Meteorological Organization (WMO), showing that the world was experiencing the warmest eight years on record, beginning in 2015 when the countries joined in the Paris Agreement (Sengupta and Gross 2022). Biden's address to world leaders at the meeting started with an apology for the U.S. pull-out of the Paris Agreement and a reminder that among his first acts as president was the return to the accord.

His comments disappointed some who supported a wide discussion of "loss and damage" or climate reparations, as Biden did not address this issue. The U.S. and other wealthy countries have long blocked calls for creation of a loss and damage fund, fearing it would make them liable for an almost unlimited amount of funding. Biden did urge all countries to cut the pollution causing global warming as the U.S. was doing. Biden also reaffirmed the U.S. commitment to provide $11 billion annually by 2024 to assist developing nations to transition to renewable energy. That commitment was part of the Obama Paris pledge and is not considered "loss and damage" funding. However, in 2022 Biden was able to get just $1 billion out of Congress for that purpose.

Biden spoke about the new climate law in the U.S., The Inflation Reduction Act, that his administration was able to pass that he said would propel the development of a cycle of innovation that would help the world confront the climate crisis. This new funding for research and development would, Biden said, help everyone in the world to achieve their transition to the clean energy future more economically. Biden also announced that for the first time the U.S. would require gas and oil companies to reduce methane leaks. Biden's comments were well-received, earning him a standing ovation (Friedman and Tankersley, Biden Casts America as Climate Leader and Promises a "Low-Carbon Future" 2022). After Biden left COP27, the U.S. did participate in talks and eventually accept loss and damage funding.

After COP27, Biden flew to Indonesia to meet with China's President Xi Jinping the day before the G20 meeting in Bali. The one-on-one talks were largely devoted to the newly emerging Cold War tone dominating the relationship of the U.S. to China including disagreements over Taiwan, North Korea, and trade. The discussion also broached the topic of climate and both countries left the meeting with an agreement to have delegates of each nation arrange subsequent meetings to further those discussions, which been frozen since August (Feng, Ruwitch and Ordonez 2022).

Congressional Dynamics During the Biden Administration

The Congress that took power with the election of Biden shared with the newly elected president the agenda of climate action. Much talk was devoted to devoted to passing a Green New Deal (GND). The GND was a plan originally drafted by youth activists. It represented a massive proposal to address climate change issues and also to restructure the economy through decarbonization and by making it more fair and just. Thus the GND embodied principles of climate action and climate justice. Once the general framework was elaborated by youth activists, politicians and others began to discuss it. The GND became a central plank in the U.S. Green Party platform beginning as early as 2008, in 2016 Bernie Sanders included it in his presidential campaign, and in 2018 Representative Alexandria Ocasio-Cortez (D-NY) included it in her run for office. She, along with her co-sponsor Senator Ed Markey (D-MA), introduced it as a Congressional resolution on February 7, 2019. The Senate voted down the resolution in March of 2019, but Ocasio-Cortez continued to push such efforts in the House (Roberts 2029).

In his run for the presidency, Biden said he did not support the Green New Deal which had come under attack by Republicans. As the politics of the first two years played out in Washington, more centrists ideas gained hold and were included in what would be passed as the infrastructure package and the climate bill. Politicians like Ocasio-Cortez and Bernie Sanders were disappointed by the extent of the efforts included but, in an attempt to gain enough support to pass legislation, compromises were made (Kurtzleben 2021). Senator Joe Manchin (D-WV) was particularly successful in getting compromises he wanted in what finally emerged as the Inflation Reduction Act, including many concessions to the oil, gas and coal industries (Plumer and Friedman 2022).

In October of 2021, Congress held hearings on disinformation flowing from the large oil companies and their involvement in duping the public about the true reality of climate change. Democrats hoped to mirror the big tobacco hearings of the 1990s and to get the oil executives to acknowledge their contribution to climate misinformation. Attending the hearings were the CEOs of Shell, BP, ExxonMobil, and Chevron. Democrats asked if the CEOs would pledge to stop lobbying against climate action to reduce GHG emissions and to promote electric vehicles. None agreed. Instead, they made statements that underscored their support for a transition to clean energy and denied that they had ever engaged in campaigns of misinformation (Tabuchi and Friedman 2021).

The Congress that served for the first two years of the Biden administration was heavily divided over many issues and climate was one of them. Not one Republican voted for what eventually emerged as the climate bill, the Inflation Reduction Act. Many Republicans in the Congress continued to deny the scientific consensus on climate change, although their approach was generally no longer outright denial but rather deflection. They deflected the issue by saying they were not scientists. These members represented the majority of the Republican caucus, and received substantial funding from the oil, gas, and coal sector. With such dynamics, efforts to win bipartisan support for a climate bill were doomed from the start (Drennen and Hardin 2021). Despite these dynamics, the Biden administration was able to eke out some legislative successes.

The 2022 mid-term elections were a success for Democrats, although they did lose control of the House of Representatives. The election was a success story in that the Biden administration did better than most presidents, historically. The president's party historically performs poorly during mid-terms but Democrats held on to the Senate and lost only nine seats in the House. Since 1932 there have been only three previous times when the president's party gained or lost no Senate seats and lost fewer than 10 House seats (1932, 1962, and 2002). In addition, the Democrats gained governorships (Enten 2022). While the loss of the House will likely result in bolder Republican actions to setback Biden's agenda or reverse legislation passed in the first two years, continued control of the Senate will considerably limit what Republicans are able to accomplish in the last years of Biden's first term.

Major Legislative Victories

The Consolidated Appropriations Act of 2020 and the American Rescue Plan

Even before Biden took office, climate legislation was beginning to appear. The bipartisan $900 trillion stimulus bill passed to avoid a government shutdown and to address the COVID-19 pandemic in December of 2020 included a number of climate provisions. The law included $2.3 trillion in tax cuts for renewable energy, and initiatives to promote carbon capture and storage. Additionally, the bill added millions to the government's science programs for 2021. It was signed into law by the lame duck Trump (King 2020).

The stimulus bill was a considerable effort to address the pandemic, but it was deemed insufficient by the Biden administration; in March of 2021, Biden signed into law the American Rescue Plan. This was a $1.9 trillion plan directed mostly at short-term pandemic relief programs; however, it also contained a few provisions to address climate. These included $30 billion for relief efforts for mass transit, which had been hit hard by low ridership in the wake of the pandemic. The bill also provided $350 billion to support state and local governments. These funds were intended to cover shortfalls in state and local funding for air and water protection, for enforcing tailpipe emissions regulations, and for implementing climate-friendly energy efficiency standards. The monies could be used for infrastructure improvements directly related to sea-level rise and local flooding (Curtis 2021).

The Infrastructure Investment and Jobs Act

The Infrastructure Investment and Jobs Act (IIJA) was signed by President Biden on November 15, 2021. It was a major bipartisan bill that included $1.2 trillion in spending on infrastructure such as roads and bridges, but it also included some climate provisions. For instance, the law provided $7.5 billion for electric vehicles including low-emission buses and ferries, and thousands of electric school buses. Another $7.5 billion was earmarked for building a nationwide network of electric charging stations. The act provided $65 billion to improve the electricity grid including thousands of miles for new transmission lines to support renewable energy sources. Another $50 billion was targeted to making the nation's water systems resilient against floods, storms, and cyberattacks (Lobosco and Luhby 2021).

The CHIPS Act

Signed into law on August 9, 2022, the Creating Helpful Incentives to Produce Semiconductors and Science Act (CHIPS) was designed to boost U.S. competitiveness by catalyzing investments in domestic chip manufacture. The Act promoted more research and development (R&D) and commercialization of leading-edge technologies. The technologies that were specified in the bill included clean energy technologies and CHIPS was authorized to spend $280 billion, the overwhelming bulk of which, $200 billion, was earmarked for R&D and commercialization. By including clean energy technology in the bill, the ability to advance clean energy technologies was achieved (Badlam et al. 2022).

The Inflation Reduction Act

Signed into law on August 15, 2022, the Inflation Reduction Act ushered in a vast new wave of spending for climate. The IRA is poorly named because it has little to do with inflation, the concern of the Federal Reserve, but offers much on clean energy, control of some drug costs, and improvement of revenue collections. Nevertheless, during a period of greater than 8 percent inflation, by passing the measures under this name, Democrats hoped to position themselves for credit should inflation decline (Newman 2022). While falling short of the initial goal outlined in the president's $2.2 trillion Build Back Better plan, the IRA was the largest single U.S. investment in stopping global warming by reducing the demand for fossil fuels, the major cause of climate change.

The IRA invested $400 billion in tax credits aimed at encouraging consumers to buy electric vehicles and pushing electric utilities toward renewable energy sources. Environmental analysis of the bill suggested that it would cut GHGs by 40 percent below 2005 levels by 2030. To get the bill through the Senate, though, Democrats agreed to a number of fossil fuel and drilling provisions to bring Senator Joe Manchin (D-WV) on board. The law assured new oil drilling leases in the Gulf of Mexico and offshore of Alaska. It also expanded tax credits for carbon capture technology that could also make existing coal and gas power plants more viable. It demanded that the Department of the Interior continue to hold new leases for oil and gas drilling on federal land if DOI approved new wind or solar ventures on public land.

More than $30 billion in tax credits were included to speed up the production of wind turbines, solar panels, and batteries. Additionally, the bill earmarked $10 billion to build facilities to manufacture electric vehicles and solar panels. Another $500 million was allocated for heat pumps. The bill also focused on environmental justice by allocating $60 billion to disadvantaged areas of the country that are harmed by climate change. The bill regulated leaks of methane and imposed high fines on oil and gas companies that do so, and reversed a 10-year moratorium on offshore wind leasing put in place by the Trump administration (Cochrane and Friedman 2022).

The IRA revealed some significant shifts in climate policy. The energy industry was largely silent about the bill, perhaps because the bill was based on subsides for clean energy rather than penalties on fossil fuel use. Holcim, a concrete and cement maker, and BP and Shell along with dozens of other companies actually signed letters promoting the bill's passage. Aside from penalties for methane leaks, the bill did not focus on limiting GHG emissions. Even though no Republicans voted for the bill, the IRA was substantially different from progressive Democratic past proposals on climate change, such as the Green New Deal (Mufson, The Surprising Political Shifts That Led to the Climate Bill's Passage 2022).

The IRA also contained some provisions to help states with their climate change efforts. The bill allocated $8.6 billion to state energy offices to help consumers make energy efficient upgrades. It earmarked $7 billion for states, cities, and tribal governments to deploy clean energy technologies. It integrated environmental justice concerns by creating a Greenhouse Gas Reduction Fund, or Green Bank, for disadvantaged communities. It gave those governments $5 billion in Climate Pollution Reduction Grants. It allocated $2.2 billion for state and private forestry conservation programs to plant trees and promote natural sinks. It awarded $1 billion for state and local governments to adopt green building energy codes, including $670 million to specifically encourage net-zero energy codes. Finally, it allocated $5 million for states to adopt tighter tailpipe emission standards (Joselow 2022).

The IRA also dealt with the assault on the administrative state launched by *West Virginia v. EPA*, in which West Virginia and its partners argued that the Supreme Court should curb the power of EPA because the federal agency had exceeded its powers as defined by Congress in the Clean Air Act which did not explicitly direct the EPA to regulate carbon dioxide (CO^2). The Clean Air Act more broadly tells the EPA to regulate pollutants that "endanger human health." In 2007, the Supreme Court in *Massachusetts v. EPA* asked the EPA to determine if CO^2 fit that description and in 2009 EPA issued a finding that said it did, marking it a pollutant that EPA could regulate.

The Obama and Biden administrations used that finding to defend regulations on tailpipe emissions from gasoline engines and coal- and gas-burning power plants. But because the court had not directly addressed the issue, court challenges continued. In *West Virginia v. EPA*, the court specifically stated that if Congress desired to move away from fossil fuels, they needed to explicitly say so. The IRA did just that. It amended the Clean Air Act to define CO^2 produced from the burning of fossil fuels as an air pollutant, leaving no doubt that Congress wanted the federal agencies to regulate CO^2. While Republicans sought to strip the language from the law, the bill passed by 51 to 50 with no Republicans voting for the measure and Vice President Harris casting the deciding vote. While inclusion of the language in the IRA does not preclude additional court challenges, it makes them more difficult to win (Friedman, Democrats Designed the Climate Law to Be a Game Changer. Here's How. 2022).

The first two years of the Biden administration included big legislative victories and represented a big step forward for the environment. About a quarter of the new climate spending was imbedded in the Infrastructure Investment and Jobs Act, so more than $100 billion in climate spending was passed with Republican support. Perhaps part of that support was gained because the legislation benefited corporations far more than individuals. But the majority of the new climate spending came from the passage of the Inflation Reduction Act, passed solely by Democrats. While it pumped more than half a trillion dollars into climate spending, it was much smaller than progressive Democrats and Biden had hoped for. The linchpin in the negotiations was Senator Joe Manchin (D-WV), who opposed the broader measures (Bhatia, Paris and Sanger-Katz 2022).

Biden and the Courts

The Supreme Court, in *West Virginia v. EPA*, curbed EPA's ability to regulate GHGs in a ruling limiting EPA's ability to restrict power plant emissions in June of 2022. The case was decided in a 6 to 3 vote, with the court's liberal justices in dissent. The ruling limited the agency's ability to regulate the energy sector, restricting it to controlling specific emission at individual power plants. The court ruled that Congress must act to give EPA the power to use wider approaches like cap-and-trade systems. The question before the justices was whether the Clean Air Act allowed EPA to issue regulations across the power sector. The court called this case a "major questions doctrine." It stemmed from Trump challenges to the Obama's Clean Power Plan, which imposed energy sector-wide solutions to GHG emissions, including the substitution of coal power plants with new sources of renewable energy.

Trump's EPA had revoked the Clean Power Plan and substituted for it a weaker plan that relaxed GHG restrictions. The Affordable Clean Energy Rule was struck down by the U.S. Court of Appeals for the District of Columbia Circuit on the last day of the Trump administration. The *West Virginia v. EPA* case was a result of Republican attorneys general coming together to weaken the ability of the executive to tackle global warming. The ruling in the case essentially said that the EPA cannot put state-level caps on carbon emissions under the Clean Air Act. That authority, they said, must come from the Congress (Liptak 2022). However, as discussed earlier, the language of the IRA included an amendment to the Clean Air Act specifically denoting carbon dioxide as a pollutant.

Intergovernmental Panel on Climate Change and Other Science Reports During the Biden Administration

In 2022, the IPCC released the last three volumes of the Sixth Assessment Report including the synthesis report, the report on mitigation, and the report on impacts, adaptations, and vulnerabilities. As had been the case with each additional assessment, the probabilities of certainty grew, and the impacts were revealed to be worse than previously thought (IPCC 2022). In addition to IPCC reports, the United Nations spokespersons continued to raise the threat of the climate emergency as an existential issue for the world (UN 2022).

The National Climate Assessment, a Congressionally mandated assessment of climate change impacts on the U.S., draft report released in November of 2022 revealed that "the things that Americans value most are at risk." The report provided detail on how climate-related disasters are becoming more common and more expensive. The report stated that the world's climate had already warmed by 1.1° C and that the situation in the U.S. was even worse. The report warned that over the last 50 years the U.S. had warmed 68 percent faster than the planet as a whole. Since 1970, the U.S. has experienced 2.5° F of warming, well above the average for the globe. Such warming creates vulnerabilities in major areas of life including threats to safe drinking water, housing security, infrastructure, farm production, health, animal species, and ecosystems. The report noted that climate-related disasters costing over a billion dollars have increased in frequency from once every four months in the 1980s to once every three weeks in 2022. The U.S. is also experiencing some of the worst sea-level rise in the world. The report makes clear that in the U.S. extreme heat is proliferating, as are drought, wildfires, and floods (Dennis, Mooney and Mufson 2022).

The Opposition During the Biden Administration

On March 18, 2021, a group of 21 states led by Texas and Montana sued the Biden administration over the Keystone XL pipeline revocation. The states argued that Biden's executive order overstepped his authority and that only Congress has the right to regulate interstate commerce (Moore 2021). While the lawsuit was pending, however, the Canadian company building the pipeline, TC Energy, officially abandoned the pipeline in June of 2021. After a decade of protests, protracted legal fights, and a series of flip-flopping executive orders, the pipeline was finally killed (Denchak and Lindwall 2022).

For years, fossil fuel producing states watched as investors pulled away from companies involved in fossil fuel production. In 2021, the state of Texas began efforts to reverse this trend. It passed a law boycotting financial firms that eschewed fossil fuel investments. Another seven states considered or followed with their own boycotts. In March of 2022, the Texas State Comptroller began sending letters to financial firms inquiring about their climate policies. The law, written by state legislator Jason Isaac, bars Texas state retirement and investment funds, worth about $330 billion, from doing business with companies that the Texas State Comptroller says are boycotting fossil fuels. The interpretation of whether a company boycotts fossil fuels is construed broadly to include firms that invest their clients' money in fossil fuels but offer green financial alternatives. Texas hired MSCI Inc., a financial consultant, to provide data on what firms it should boycott, despite the fact that MSCI Inc. itself has committed to carbon neutrality before 2040 (Ariza and Buchele 2022).

In a similar fashion, West Virginia announced that several major banks including Goldman Sachs, JPMorgan, BlackRock, Morgan Stanley, and Wells Fargo would be barred from government contracts within the state because they are reducing investment in coal. West Virginia, Louisiana, and Arkansas pulled more than $700 million out of Black-Rock, the world's largest investment manager, over complaints that BlackRock focuses too heavily on environmental issues. Pennsylvania, Arizona, and Oklahoma joined in an effort to stop the nomination of federal regulators that want to require that banks, funds, and companies disclose their financial risks posed by climate change.

These efforts, spearheaded by the State Financial Officers Foundation, a nonprofit organization, began with the election of Biden. The Foundation is pushing Republican state treasurers to use their power to promote oil, gas, and coal. Working with the State Financial Officers Foundation are the Heritage Foundation, the Heartland Institute, and the American Petroleum Institute. At the same time as major banks and corporations attended the global climate summit in Glasgow, the Republican state treasurers and State Financial Officers Foundation held a conference in Orlando, Florida, to discuss how to stop green efforts from going forward. The Texas legislation is seen as a model to be copied and passed in other fossil fuel producing states (Gelles 2022).

Regional, State, and Local Action

Regional and multi-state efforts continued into the Biden administration. The Regional Greenhouse Gas Initiative (RGGI) remained active and was strengthened in 2020 when New Jersey returned to the group. The group will be further strengthened by the likely addition of Pennsylvania in 2022, becoming the 12th state to be a partner in the mandatory cap-and-trade program limiting CO^2 from the energy sector. However, opposition within the state may slow or stop Pennsylvania's entry. Virginia in 2020, passed the Virginia Clean Economy Act which instructed the state's Air Pollution Control Board to put in place a cap-and-trade program following RGGI's model rule (RGGI 2022). The Western Climate Initiative (WCI) continued its existence in 2022 with two members, California and the Province of Quebec. Both have established cap-and-trade programs, and both committed to forming one trading market covering both regions.

The U.S. Climate Alliance, formed after Trump pulled the U.S. out of the Paris Agreement, grew to 25 states and Puerto Rico. Member states are committed to reducing their GHG emissions consistent with the Paris Agreement. The Pacific Coast Collaborative, a cooperative agreement between Alaska, British Columbia, California, Oregon, and Washington, continued in 2022 to pursue a series of joint efforts on infrastructure, energy, climate policy, and transportation. The Transportation and Climate Initiative (TCI), a cooperative agreement between mid–Atlantic states, was expected to launch in 2022 (C2ES 2022).

California remained a key player in the push for climate action during the Biden administration. In August of 2022, it made history: The California Air Resources Board approved a broad plan to ban the sale of new gasoline-powered cars and light trucks by 2035. By that date, all new cars sold in California must be electric vehicles or other zero-emissions models. This move is important because California is the largest automobile market in the U.S. and because many states follow California's lead (Karlamangla 2022).

Conclusion

The Biden administration ushered in a broad swing back to climate crisis action. This could be seen through the appointments of climate crisis activists to his government and his early executive actions to address climate change. Within hours of assuming the presidency, Biden swiftly reversed much of what Trump had done through his executive actions.

The Biden administration, in its first two years, was also tremendously successful in passing new legislation to promote climate action. This is important because, once passed, only a Congressional reversal or the courts via challenges can change the policy's direction. Biden's legislative approach was different from that used earlier, focusing on incentives rather than regulations to achieve the desired goals. This approach aligned with the Biden administration's desire to bring the private sector into the fight against climate change as a partner rather than as an adversary. While this legislation fell short of what the more progressive coalition of Democrats wanted, it still represented a significant step forward for climate action.

Biden's presence at international meetings returned the U.S. to a position of climate leadership worldwide that had been greatly tarnished by the Trump administration. However, even U.S. allies adopted a tentative acceptance of the U.S. leadership role given the political division in America that could, with one more election, return a conservative Republican to the White House. The ongoing Cold War–like hostilities between the U.S. and China also raise questions about how influential the U.S. might be. Garnering Chinese cooperation on climate actions will be more difficult in an era of conflict with China over other issues.

The mid-term election results changed the situation for the Biden administration going into the last two years of his first term. In one way, the mid-terms strengthened Biden because the Democrats did not experience the normal losses a standing president typically does. Biden's hand, in that regard, was strengthened, both domestically and internationally. But the taking of the House by the Republicans brought a new domestic reality into the mix. Republicans, in the last part of Biden's first term, will hold the House by a fragile majority within a party suffering from factional differences. This will weaken Republican efforts to undermine Biden's agenda, but it will also mean that in the last part of Biden's first term, progress toward strengthening the climate agenda will stall. The remainder of the first term will be dedicated to implementation of the legislation passed in the first two years and to trying to hold Republicans' efforts to reverse those advances at bay.

CHAPTER 8

The Centrality of
the Climate Justice Movement
to 21st-Century Activism

Introduction

Climate justice has evolved from an earlier form of environmental discontent, typically called environmental justice or, alternatively, environmental injustice or environmental racism. Whatever phrasing is used to identify it, environmental justice is rooted in the notion that people of color are disproportionately exposed to the problems of pollution and environmental degradation compared to members of the majoritarian population. In the United States, where the movement began, this fundamental injustice was largely associated with race, although race being highly correlated with socio-economic status (SES) brought poor white people into the mix of those adversely affected. The idea of environmental justice spread worldwide in the last decades of the 20th century. As concerns over the climate crisis grew to be the existential threat for the world's environment, the environmental justice movement expanded to include climate justice.

The climate justice movement argues that poorer nations and populations will be made to bear more of the burden of climate change impacts and that they will have fewer resources to adapt to the climate crisis. For example, in 2022 alone there was severe drought in Somalia, which led to the displacement of more than one million people. The South American states of Uruguay, Brazil, Paraguay, and Argentina experienced a heat wave that heavily impacted gain and other crop harvests. Billions of people in India suffered from severe heat without the refuge of air conditioning. In eleven Sub-Saharan African countries, tension between herdsmen who move their cattle from one place to another and sedentary farmers erupted because of changes in rainfall patterns. These groups traditionally lived together in peace, but climate changes are now provoking conflict. Pakistan was heavily flooded by a monsoon, leaving one third of its land submerged and more than 1,000 dead. Island nations fear the loss of their entire countries to sea-level rise. These are just a sampling of the climate crises affecting the developing world.

The moral response to this is the demand that rich nations transfer money and other resources to poor nations so that they can more adequately deal with the ravages of climate change. In addition to this claim, the climate justice movement also embraces the idea of intergenerational injustice. The contention is that the older people in the world are passing to the next generation the burdens of climate change. The older generation experienced the benefit of the burning of fossil fuels but the younger generation will experience the negative consequences of these actions. As such, climate justice became a fundamental tenant of the youth climate change activist movement that arose in the second decade of the 21st century.

This chapter explores both of these currents and discusses how they impact the national and international approaches to the climate crisis. It begins with a brief overview of the history of the expansion of environmental justice movement to include climate justice issues. The chapter explores how this played out with the growing international and domestic youth movements and in the demands by poorer countries that something be done about their plight.

The Evolution of the Climate Justice Movement

In the U.S. the environmental justice movement began in the 1971 when the Council on Environmental Quality (CEQ) raised the issues of race and SES being associated with environmental risk. Soon social and religious organizations got involved. The United Church of Christ (UCC) brought widespread attention to the issue in 1987 when its UCC Commission for Racial Justice published *Toxic Waste and Race*. The report focused on the correlation between race and the likelihood of living near a toxic waste site. The report got widespread consideration with its allegations of racism and the untoward impacts of pollution on people of color and the disadvantaged. The movement grew to national attention in 1991 with the gathering in the nation's capital of the National Peoples of Color Environmental Leadership Summit.

The main position of the movement was that people of color and of lower SES were disproportionately exposed to environmental health and safety risks, especially those associated with toxic chemicals. Taking these concerns seriously, the Environmental Protection Agency (EPA) established the Office of Environmental Justice in 1992. In 1994, President Clinton issued executive order 12898 instructing all federal agencies to ensure that their programs did not inflict environmental harm on the poor or minority communities. Robert Bullard's books *Dumping in Dixie: Race, Class, and Environmental Quality* and *Confronting Environmental Racism: Voices from the Grassroots* both released in the 1990s further documented the issue (D. Rahm 2019).

Since its inception, the environmental justice movement grew in size and influence. By the end of the first decade of the 21st century, it included many thousands of organizations. In addition, the ideas of the environmental justice movement were actively adopted by other established environmental and social organizations which added environmental justice to their agendas. An increasing number of community groups, university centers, legal clinics, faith-based groups, labor, and youth groups formed partnerships and collaborations to work with people of color and lower SES to address the issues of environmental safety and health risk that affect them. The rise of the internet provided a relatively costless way for these organizations to network (Rosenbaum 2020).

The environmental justice movement also became international in the aftermath of its start in the U.S. Globally, advocates for human rights, women's rights, and indigenous rights adopted the dialog of the U.S. environmental justice movement in their grassroots struggles over equitable access to land, water, food and energy. In the international framework, environmental justice calls for the equitable distribution of the burdens and benefits of economic activity, procedural rights that guarantee that members of all communities have a say in environmental issues that concern them, and the corrective justice that ensures compensation for those who are harmed by the actions of others. In the global context, environmental justice cannot be separated from other forms of social and economic justice, and advocates argue it cannot be achieved without addressing the wider problems of poverty and racism (Gonzalez and Atapatta 2017).

The international environmental justice movement began with an emphasis on the environmental conflicts that exist between rich and poor nations because of the distributive injustice that has resulted from the most affluent populations of the world consuming the vast majority of the world's economic output. Moreover, the rich nations generate the overwhelming bulk of all hazardous waste, which all too often is exported to the developing countries of the world for disposal. An element of procedural justice also plays out in that the rich nations often dominate discussions and decisions in international organizations such as the World Bank, the International Monetary Fund, and the World Trade Organization. In such discussions and decisions, the concerns of developing nations are often downplayed or ignored. In the international context, the environmental justice movement also focuses on the colonial and post-colonial periods in which the developed nations grew wealthy by appropriating the resources of the developing nations, thus exacerbating the problem of poverty (Gonzalez and Atapatta 2017).

As climate change became the most concerning environmental issue, these worries extended over to the new issue. With the problem of climate change, most nations acknowledge that rich developed countries made their wealth in large part through the burning of fossil fuels in their early industrializations. These harmful emissions will cause the most damage to the developing world, which itself has emitted only a small portion of the pollution. These poorer nations have still not industrialized and struggle to develop at the same time. Their poverty makes them unable to use adaptation solutions to build resilience without financial and technical help from the developed world (Shue 2014).

The broader arguments elaborated by the developing nations as well as international environmental and climate justice advocates, caught on with many climate change advocates in the affluent developed countries. Climate change became increasingly understood as intertwined with the issues of justice and equity. Particularly sensitive to these moral arguments were the newly emerging youth activists who picked up on these ideas and began to echo them in their own purposes and statements. In addition, the youth movement introduced another important consideration into the climate justice debate—intergenerational injustice. This argument is rooted in the fact that the next generation will inherit a world made worse by climate change and that present generations have an ethical and moral obligation to future generations. Questions such as what risks the current generation has a right to impose on future generations and how available resources can be used currently so as to not deny their use to future generations arise (Schuppert 2012).

The climate justice movement, therefore, developed to represent a far broader set of concerns than those originally part of the core issues elaborated in the earlier environmental justice movement. The key activists in this movement included those most likely to be adversely affected by climate change impacts—the youth and the developing world.

The Youth Movement

While the participation of youth in political movements is historically not uncommon, youth have played and continue to play an outsized role in the climate crisis debate. Much of this is owed to the fact that youth see their futures being directly harmed by the lack of progress nations have made in the fight against climate change. Recognizing this, the United Nations began to include youth in activities sponsored by the United Nations Framework Convention on Climate Change (UNFCCC) in the early 21st century (discussed below). Widespread grassroots youth activism also emerged with the goal of bringing attention to the climate crisis and forcing action on political entities. This youth grassroots movement initially drew heavily from the very young, in many cases teenagers. As the youth movement endured and grew, the early members aged somewhat and it also attracted older sympathizers. As with the environmental justice movement, traditional environmental groups and advocates added intergenerational climate injustice to their agendas, as well, thus adding allies to the movement.

Birth of the Youth Movement

In the U.S., the Sunrise Movement launched in 2017. The Sunrise Movement is a "youth movement to stop climate change and create millions of good jobs in the process" that claims to be building an "army of young people to make climate change an urgent priority across America, end the corrupting influence of fossil fuel executives on our politics, and elect leaders who stand up for the health and wellbeing of all people" (Sunrise Movement 2022). The movement first launched in 2017 with the goals of making climate change an urgent issue in America and to remake climate policy under the banner of the Green New Deal. The Green New Deal was the platform of Democratic progressives to not only fight climate change but also do so by creating many new well-paying jobs in the clean energy economy. It included as one of its premises that confronting the climate crisis required simultaneous confrontation of race and poverty in America, thereby linking it to the larger justice movement.

In 2021, those in the Sunrise Movement refined this plan, since the original plan was intended for four years only and they believed their original strategy would not work given the new reality of the Biden administration and the Covid-19 pandemic. They heavily engaged in the 2018, 2020, and 2022 elections, supporting "bold transformative climate legislation, for abortion rights, for immigration, for gun control...."

In 2022, the movement had more than 400 local hubs representing local Sunrise chapters across the country (Sunrise Movement 2022).

The Justice Democrats, a political action committee (PAC) supported by grassroots donations, formed to help progressive candidates. Their goal is to create a mission-driven caucus in Congress to push for more progressive solutions to the nation's most pressing problems. Their platform includes three planks. First, an economy for all that supports cancellation of student debt, the Green New Deal, Medicare for all, free public college and trade school, expanded Social Security, and securing a living wage. Second, a society for all that includes reparations for the injustices of slavery and its aftermath, housing as a human right, disability justice, reproductive rights, prison reform, LBGTQ+ equality, and the end of gun violence. Third, democracy for all underpinned by voting rights, immigration rights, democratic reform, and a progressive foreign policy (Justice Democrats 2022). While not restricted to youth, the Justice Democrats represent the younger more progressive Democrats, many of whom also affiliated with the Sunrise Movement.

Just as many in the Justice Democrats were connected to the Sunrise Movement, so too are many in the New Consensus. New Consensus is a small think tank dedicated to providing detailed plans for how the government could implement programs to make it possible for everyone to live safe, comfortable, and sustainable lives. New Consensus is made up of a different generation of thinkers exploring how government and other public institutions can transition to a green economy, close the wealth and income gaps, and create new high-wage jobs (New Consensus 2022).

While the Sunrise Movement, New Consensus, and Justice Democrats were largely domestic U.S. groups, early in the building of the youth movement an international component was launched. On August 20, 2018, Greta Thunberg, a Swedish teenager, called the first student climate strike. Thunberg had been inspired by a school walkout against gun violence that had happened in Parkland, Florida, in response to a school shooting there. Thunberg, a 15 year old, decided to spend her Fridays not in school but sitting in front of the Swedish parliament in protest of its lack of action on the climate crisis.

With this modest beginning a movement called Fridays for Future was born. Fridays for Future, which supports Friday school strikes to protest the climate crisis and the lack of action on it, spread across the world. Its appeal led to, in March of 2019, the first global strike which drew more than one million people together for the strike. In September of 2019, young people (with support of many adults) joined a call issued by Thunberg and other activists to strike in an event called Global Week for Future in the ramp up to the 2019 UN Climate Action Summit (Fisher and Nasrin 2021).

The U.S. Youth Climate Strike movement was started in January 2019 in solidarity with Greta Thunberg's Fridays for Future. It was organized to push for climate action in the U.S., as well as to hold training and workshops on climate justice, and youth climate activist training (U.S. Youth Climate Strike 2020). Fridays for Future U.S. was formed as part of the worldwide Fridays for Future movement. It is a grassroots, decentralized movement spanning the U.S. with over 40 local chapters. They advertise that they are school children fighting for a livable future (Fridays for Future U.S. 2022).

Schools for Climate Action is a U.S.-based nonpartisan, youth-adult led grassroots organization with the mission of getting schools to fight climate change. It helps school boards, PTAs, school environmental clubs, teachers' unions, and school support organizations to pass climate action resolutions. It supports youth being an integral part of the drafting of climate actions, providing school boards with a model climate action resolution from which to begin their discussions. The organization formed in 2019 after the climate change-fueled Tubbs Fire destroyed schools in Sonoma County, California. In the aftermath of the fire, more than a dozen Sonoma school boards passed climate resolutions that spoke up for their students' futures in a climate impacted world. The group contends that climate change is a generational justice issue and also pressures teachers' pension funds to divest from fossil fuels (Schools for Climate Action 2022).

This Is Zero Hour is a youth-based organization begun in 2017 with the goal of organizing a national day of protest led by youth. Zero Hour's platform includes demands for climate justice that includes racial justice, economic justice, and equity (This Is Zero Hour 2022). Zero Hour filed a lawsuit through Our Children's Trust against the federal government. Our Children's Trust is a nonprofit public interest law firm that specializes in securing the legal rights of youth to a safe climate (Our Children's Trust 2022). They filed *Juliana v. United States* in 2015. It represented 21 youth who argued that the federal government's inaction on climate change had unconstitutionally violated the rights of the youngest generation to life, liberty, and property because of the federal government's policy of permitting and encouraging the combustion of fossil fuels. As of 2022, *Juliana v. U.S.* remained unresolved and continued to be litigated (Our Children's Trust +Youth v. Gov 2022).

U.S.-based groups that had gotten their start focusing only on youth joined over time with more established environmental groups with broader demographic appeal. Two of the most prominent U.S.-

established groups include 350.org and Sierra Club but youth activists are also joining UK-based Extinction Rebellion and Just Stop Oil (Fisher and Nasrin 2021). Extinction Rebellion focuses on two crises in tandem—the climate emergency and the extinction crisis—arguing that the answer to these is ecological justice (Extinction Rebellion 2022).

Just Stop Oil, organized in 2022, seeks coalitions with other groups as well, although its tactics, being quite aggressive, might preclude coalitions with less radical groups (Just Stop Oil 2022). Just Stop Oil, fueled by anger, engages in actions to disrupt everyday life in the UK including stopping traffic on highways around London. In addition, Just Stop Oil engages in protests to draw public attention, such as throwing tomato soup at artwork and gluing themselves to the walls of museums. Police in the UK have been empowered to be more aggressive with the protesters, raising some concern that the freedom of speech rights of protesters would be violated (Castle 2022).

A group of university friends that included Bill McKibben, author of *The End of Nature*, founded 350.org in 2008. Its stated goal was to build a global climate movement. Over time 350.org became a global organization. Like most organizations fighting climate change, 350.org sees its mission as fighting for justice in its attempts to strengthen the voices of communities most hard hit by the ravages of climate change. It supports a rapid and equitable shift to renewable energy and an immediate ban to all fossil fuel projects everywhere (350.org 2022). Sierra Club, one of the oldest environmental organizations in the U.S., added climate change to its list of environmental causes as the problem emerged (Rosenbaum 2020).

Youth Leadership in the Climate Crisis

Greta Thunberg has had an outsized role in climate crisis events and meetings, and although other youth have played widespread roles, Thunberg became the symbol of the youth movement. Thunberg first became a voice with her 2018 school strike. She spoke at nearly every COP held since 2018 with the same basic message, that governments and world leaders were not doing enough to stop global warming. She traveled to the U.S. to attend the UN Climate Action Summit but the week before that appearance, she testified before Congress. She submitted the Intergovernmental Panel on Climate Change's (IPCC) latest report as her written testimony because she said she wanted Congress to listen to the scientists. She spoke to two committees—the House Foreign Affairs Subcommittee on Europe, Eurasia, Energy and the Environment and the House Select Committee on the Climate Crisis. Three other youth activists joined her, including Jamie Margolin, Vic Barrett, and Benji Backer. She also attended the protest outside the White House and met with former President Barack Obama. Thunberg gained widespread attention because she traveled to the U.S. on a zero-emissions boat (Hayes 2019). On September 23, 2019, she spoke at the UN Climate Action Summit in New York. Her message to world leaders was provocative. She accused the leaders of not doing their jobs and ignoring years of scientific advice. She claimed that the world's leaders had stolen her childhood and dreams with their empty words. She claimed that the eyes of all youth are on world leaders who have failed to provide a livable future for the next generation (NPR 2019).

On Earth Day, April 14, 2021, Thunberg again testified before the House Oversight Committee. In this testimony she criticized fossil fuel subsidies, saying that future generations would hold the current generation accountable for their actions (Adragna 2021). That September she spoke at the Milan pre–COP26 summit for youth summing up her climate justice arguments saying: "And the climate crisis is of course only a symptom of a much larger crisis—the sustainability crisis, the social crisis—a crisis of inequality that dates back to colonialism and beyond—a crisis based on the idea that some people are worth more than others and therefore have a right to exploit and steal other people's land and resources—and it is very naïve to believe that we can solve this crisis without confronting the roots of it." Also included in her comments was the accusation of intergenerational injustice—"We can no longer let people in power decide what hope is. Hope is not passive. Hope is not blah blah blah. Hope is telling the truth. Hope is taking action" (Thunberg 2021).

Inclusion of Youth in International Discussions

Youth have had a place in climate negotiations since 2005. In that year YOUNGO (the official youth constituency of the UNFCC) was formed by the United Nations Framework Convention on Climate Change secretariat; it is the official youth-constituency to the UNFCCC. In 2005, it organized the first Conference of Youth (COY) to take place right before the annual conference of parties (COP). The COY's objectives included providing capacity building and policy training to prepare youth for engagement in the upcoming COP, sharing knowledge and experience, and building youth networks and movements. The first COY took place before COP11 in Montreal and has continued at every COP since. The purpose of the COYs is to assure that the voice of youth is considered in the COP meeting. YOUNGO organizes COYs in different ways to assure access. These include the Global COY which is the official in-person meeting held before the COP, a virtual COY (vCOY) so that those youth who cannot travel can participate, as well as regional and local COYS (United Nations Climate Change 2022).

The UNFCCC also provides an Action for Climate Empowerment (ACE) group for youth which focuses on six themes: education, public awareness, training, public participation, public access to information, and international cooperation. Organizing youth to participate in youth-related events at the COPs, ACE launched a Hub in 2022 to increase public support for, and engagement in, climate activities that can help accelerate the implementation of the Paris Agreement. Youth are a special focus of the ACE Hub. The ACE Hub provides an annual forum called the ACE Youth Exchange and Hackathon to develop new ideas to implement climate action (United Nations Climate Change 2022).

In 2020, the United Nations announced the formation of the Youth Advisory Group on Climate Change. The seven-member group, made up of youth climate leaders between the ages of 18 and 28, was formed to provide direct advice to the Secretary-General of the United Nations. The group includes young activists from Brazil, Fiji, France, India, Moldova, Sudan, and the U.S. While the full impact of the group not yet known, its establishment marks the first time youth group will have a formal role in advising the Secretary-General, making it clear that the UN considers the input of youth essential (Herr 2020).

Climate Anxiety

One of the impacts of the injustice of the intergenerational consequences of the climate crisis has been the impacts for the health and futures of children and young people. The world they will inherit will be ravaged by climate change and yet they have little ability to limit the harm. This renders them vulnerable to climate anxiety or eco-anxiety. Because climate anxiety is rooted in rationality, it does not imply mental illness. Rather, it may have the beneficial effect of leading people to reconsider their actions in order to respond appropriately to the threat that is climate change. However, as a complex phenomenon, it can also lead to grief, worry, and fear that are tied to anticipation of loss. Substantial levels of climate-related distress have been reported globally, with youth and children particularly susceptible. Parents, educators, and clinicians have reported hearing great concern about climate change from youth. This is troubling because exposure to chronic stress in youth can lead to problems in later life. Children and youth often feel a sense of abandonment, confusion, and betrayal because of adult inaction on climate change (Hickman et al. 2021).

The climate mental health crisis takes its toll on those who have lost much in the worsening climate crisis. It has been seen in farmers in Australia, India, and elsewhere who face increasing challenges growing food in a rapidly changing climate. It is also hitting indigenous and vulnerable communities, for whom the climate crisis is just the latest in the decades or centuries of social oppression. It impacts parents, who feel they cannot protect their children. And it is impacting youth and children who see their futures in peril. Youth, in particular, are prone to reporting that they see the future as frightening and that makes them feel sad and anxious. Perhaps the situation can be illustrated by the caption on a youth protester's sign in a climate school strike that said, "Why am I studying for a future I won't have?" (Kalmus 2021).

Developing Countries Versus Developed Nations

The second major group that plays a large role in the climate justice debate is the coalition of developing countries who are being impacted by the ravages of climate change as they endure floods, severe storms, heat waves, drought, rising sea-levels and other climate-related disasters. While all countries are experiencing such disasters with increased frequency, developing countries are the least able to build their resilience simply because of their level of poverty. The injustice aspect of this situation comes from the fact that these developing countries contributed to the pollution of the atmosphere least of all nations. The developed countries became wealthy largely by creating the pollution now driving climate change but their wealth enables them to adapt and build resilience.

History of Demands at International Conferences for Assistance

Since the start of global meetings on the environment, there has been a clear demarcation line drawn between the affluent developed nations and the poor developing nations. This was evidenced in the very first global meeting in 1972, with Proclamation 4 of its Stockholm Declaration. This proclamation established what became known as the concept of differentiated responsibility. It stated that for the affluent developed nations, the environmental problems of pollution stemmed from their industries and so developed nations should seek to control pollution. However, for the poor developing nations, proclamation 4 stated that the environmental problems they created came largely from their poverty and failure to develop.

Poverty limits choices, often forcing marginal populations to use resources in ways that are less than optimal and that results in environmental degradation. Therefore, developing nations, according to the Stockholm Declaration, should focus first on developing and only worry about pollution after they have moved out of poverty. The Stockholm Declaration further elaborated 21 principles of which principle 9 further advanced this thinking by saying that accelerated development would be most likely to take place if the developing countries received technology and finance assistance from the developed nations (Stockholm Declaration 1972). Thus, the Stockholm Declaration elaborated the moral and ethical framework for the world moving forward to solve the globally shared problem of the environment.

Twenty years later at the Rio Earth Summit, the Rio Declaration mirrored the framework put forth in Stockholm. Principle 7 more fully elaborated the meaning of "common but differentiated responsibilities" by stating that the developed nations acknowledge the responsibility they bear for growing wealthy while contributing to the world's environmental problems. It further acknowledged their wealth and technological prowess. Principle 3 also acknowledged the right of developing countries to develop while Principle 6 acknowledged the special needs of developing countries (United Nations 1993).

Hence, by the time of the negotiations for the first binding treaty on climate change, the Kyoto Protocol, several ideas were firmly in place. First, that there were basically two kinds of countries, developed and developing, and, second, that each group had "common but differentiated responsibilities." The responsibility of the developing world was to develop first and only secondarily direct their efforts to pollution abatement, while the affluent developed countries needed to immediately reduce pollution and assist the developing nations to develop.

This moral and ethical structure was adopted by the Kyoto Protocol, which bisected the world into the two groups and demanded nothing from developing countries. While European countries accepted this framework, it was rejected by the United States. The consequences of this rejection did much to reduce the effectiveness of the Kyoto Protocol and would serve as the primary negotiating hurdle for a successor treaty (D. Rahm 2010).

The Justice Root of the Problem

Two matters arose in sorting through this issue. The first had to do with the changing position of China and other low-income nations. When the original concept of "common but differentiated responsibilities" was formed, China was clearly a poor developing country with a per capita GDP of $131 USD. By the turn of the century, though, China's per capita GDP had grown to $959 USD and by 2021 to $12,556. This compares to comparable U.S. figures of $5,609 in 1972, $36,330 in 2000, and $69,287 in 2021. In 2021 the per capita income of the European Union countries was $38,234 USD. So, while China has gotten much wealthier, it still, on a per capita basis, is poorer than the U.S. While at the same time, China has grown wealthier than all other developing nations. For instance, India's per capita GDP in 2021 was $2,277 USD compared to Brazil with a per capita income of $7,518 USD (World Bank 2021).

China is particularly important in this discussion because on an annual basis, China went from being a low emitter of greenhouse gases (GHGs) to being the world's largest single emitter. These measurements are typically done in carbon dioxide equivalent (CO_2e) amounts which adjusts substances for their warming effects and puts them all in one number. For instance, China emitted 2.89 billion tons CO_2e in 1990 as compared to 12.06 billion tons CO_2e in 2019. By comparison, the U.S. emitted 5.42 billion tons CO_2e in 1990 but 5.77 billion tons in 2019. This was down from 6.37 billion tons emitted in both 2000 and 2007. The economic downturn of the pandemic reduced world emissions somewhat. For the top six emitters in 2021, China emitted 11.47 billion tons CO_2e, the U.S. 5.01 billion tons, India 2.71 billion tons, Germany 674.75 million tons, Brazil 488.88 million tons, and France 305.96 million tons (Ritchie and Roser 2022).

Much of the developing world has a much shorter history of CO_2 emissions than the developed world. Since CO_2 stays in the atmosphere for many hundreds of years, even annual contributions going back a very long way are still influencing today's climate. If one considers cumulative contributions of GHGs to the atmosphere, a different story emerges. Since 1751, near the beginning of the Industrial Revolution, the U.S. has been the largest contributor, accounting for 25 percent of world emissions. The EU-28 was not far behind with responsibility for 22 percent. China ranks next with a cumulative contribution of 12.7 percent. Following in order of contribution are Russia (6%), Japan (4%), India (3%), Canada (2%), South Africa (1.3%), Mexico (1.2%), Ukraine (1.2%), and Australia (1.1%). All other nations in the world have contributed 1 percent or less (H. Ritchie 2019).

The heart of the justice issue is that poorer nations and those with a shorter history of development are not the cause of the problem but will likely suffer the consequences more than those that caused it because they have fewer resources to adapt to living in a warmer world and its devastating climate outcomes. For decades the developing world's nations have demanded climate justice. But also part of the story is the reality of the current situation. For the world to stay below the commitment made in Paris of 1.5°C, *all* countries must cut their current emissions. The Paris Agreement made possible a solution to this dilemma although the determination of the developing nations to receive climate loss and damage payments remains a consistent demand.

The Paris Agreement and Subsequent COPs

As discussed earlier, the Paris Agreement was fashioned to end at least some of the log jam that was preventing all nations from fully participating in CO^2 reductions. By relying on a system of nationally determined contributions (NDC), the Paris Agreement was successful in bringing all countries into the process of GHG reductions. However, in the years following the 2015 agreement, few countries had lived up to their pledges. For instance, while 32 percent of countries committed to making a 50 percent decrease in emissions by 2030, only 28 percent had submitted a detailed plan describing how they planned to meet that goal. And while 40 percent of nations agreed to become carbon neutral by 2050, only 12 percent of countries had submitted detailed plans on how they intended to accomplish that goal. The stubborn fact remains that the country pledges made in 2015 are not enough to reach the goal of limiting global temperature increase to 1.5° C by the end of the century. All nations need to increase their pledges (Climate Scorecard 2021).

The Marrakech, Morocco, COP22 held in 2016 was widely seen as the conference at which countries could showcase their progress in updating their 2015 pledges and begin making progress on reaching the goal of 1.5° C by the end of the century. When the 200 delegates arrived at the conference, it was widely believed that Hillary Clinton would be elected as the next president of the U.S. but on the third day of the conference, Donald Trump was elected. This threw the conference into disarray as delegates wondered if he would fulfill his promise to withdraw the U.S. from the Paris Agreement.

In terms of climate justice, few agenda items moved forward. Finance remained a sticking point with little progress made toward firm plans to reach the Paris Agreement level of $100 billion a year by 2020 to help developing nations move forward with mitigation and adaptation. At Marrakech, countries did approve a five-year working plan on "loss and damage" that was to begin the formal process of countries discussing topics such as the slow-impact onset of climate impacts, noneconomic losses such as culture and identity, and migration. Loss and damage concerns therefore were framed as problems nations will experience that are beyond their adaptation efforts (Yeo 2016).

Climate justice issues were at the forefront of COP23 held in Bonn, Germany, in 2017 because for the first time the conference was presided over by the prime minister of a small-island developing nation, Fiji. Climate justice issues also loomed large as a split between developed and developing nations emerged. China and India asked for an agenda item to be added to address the emission cuts developed nations were required to make prior to 2020 to be in compliance with the Kyoto Protocol. Developed nations blocked this and instead all parties agreed to provide additional information on emissions reductions in 2018 and 2019 and additional information on finance in 2018 and 2020. The countries also agreed to hold discussions on loss and damage in 2019. These talks were to focus on providing support for the victims of climate change. Parties also finalized the Gender Action Plan as well as the Indigenous People's Platform. Both have the aim of increasing the participation of traditionally marginalized groups at UN meetings (United Nations 2017).

The COP24, held in Katowice, Poland, in 2018, had the main goal of finishing the "rulebook" for the Paris Agreement, which was to enter into force in 2020. The "rulebook" is meant to define the role of carbon markets and the forms of international agreements for the Paris Agreement to go into effect. The meeting was able to hash out an agreement that every nation will report its emissions and progress in cutting them every two years beginning in 2024. Loss and damage, caused by unavoidable impacts of climate change, was added to the rulebook but not to the satisfaction of many developing nations (Evans and Timperley 2018).

Finalization of the "rulebook" was scheduled to take place at COP25 held in Madrid, Spain, in 2019. The Madrid conference was marked by protest against the countries not making sufficient progress to reach the Paris Agreement's goal of 1.5° C. The conference made clear that the sum of the NDCs for 2030 show that the parties will be 38 percent too high to meet that goal. These protesters were joined by youth activist Greta Thunberg, who spoke at the conference. The loss and damage talks at Madrid fell short of expectations of the developing nations, who suffer and will continue to suffer irreversible and non-adaptable impacts of climate change at a much higher rate than developed nations.

Climate finance in general proved a difficult issue with developing nations saying that developed nations fail to provide the financing they need. Loss and damage was particularly an issue raised at a press conference called by the Alliance of Small Island States (AOSIS) in which they argued that they needed clear and predictable finance to provide them real compensation for losses and damages that so many of the small island nations are experiencing. The G77 and China group also released a document calling for "adequate, easily accessible, scaled up, new and additional, predictable finance, technology and capacity building." Vulnerable nations called for additional funding for the Green Climate Fund for loss and damage. Pressure by the developed nations, led by the U.S., resulted in no agreement. The final text released only urged developed nations to scale up finance (Evans and Gabbatiss, COP25: Key Outcomes Agreed at the UN Climate Talks in Madrid 2019).

The next major talks were held at COP26 in Glasgow, Scotland, in 2021. The talks had been postponed from 2020 due to the pandemic. The Glasgow meeting reaffirmed nations' resolve to reach the 1.5° C goal of the Paris Agreement, although they acknowledged the gap and pledged to work toward improved NDCs that would reach that goal. Nations again agreed to provide the $100 billion in funding to support the efforts of developing nations. The rulebook was finalized, and it included provisions for loss and damage (United Nations 2021).

The new agreement that came out of the summit, the Glasgow Climate Pact, set the agenda for the next decade. It was agreed that nations would meet in 2022 to further cut emissions. Pledges at the time of the Glasgow meeting would allow for warming of 2.4° C so to meet the goal of 1.5° C, additional cuts must be made. For the first time at a COP, there was an explicit agreement to cut coal use, however, countries only agreed to the weaker goal of "phasing down" rather than "phasing out" coal use. Once again, the affluent nations promised to help poor countries with their adaptation and making the shift to clean energy. There was also discussion of a potential trillion-dollar fund beginning in 2025 which pleased some African and Latin American countries but was too little for others. At Glasgow, the U.S. and China agreed to cooperate more over the decade in areas including methane reduction and the switch to clean energy (BBC News 2021).

The issue of climate reparations was one of the main topics of discussion at COP27 in Sharm el-Sheikh, Egypt, in 2022. The developing countries demanded action on loss and damage that had been resisted by developed nations in prior meetings. It was feared that the COP talks could collapse over questions of finance. The two-week meeting was extended to allow additional discussion. Vulnerable nations, led by Pakistan, argued that developed nations owed them money because they are suffering the worst effects of climate change that has been caused by the cumulative emissions of developed nations. Pakistan suffered devastating floods in 2022 that covered more than one third of the nation and killed more than 1,200 people, and Pakistan demanded compensation for damages. The demand was opposed by the U.S., which took the position at the beginning of the talks that there would be no new fund, fearing it would be a blank check for future liabilities. The U.S. subsequently went silent in negotiations on this issue but the European Union finally proposed a special fund that would be supported by a "broad donor base" that might include China, since it is the third largest emitter of cumulative GHGs.

China is typically grouped with the developing nations, but this EU proposal moved China into another position (Rannard 2022). In the end, though, China balked at providing funding for loss and damage. The U.S. and the EU finally agreed to the loss and damage fund to help for loss and damages of resources after it was clarified that it does not contain any liability or compensation provisions. Details of how the fund would operate were left incomplete at the close of COP27; nevertheless, establishment of the fund was seen as a major climate justice victory for developing nations (Kaplan 2022).

Conclusion

Climate justice demands stem from two sources—the youth movement and the coalition of developing nations. The youth movement sees the failures to deal with the climate crisis as intergenerational injustice as the youth will inherit a world devastated by the climate crisis. The developing world sees the injustice of the poor nations having to deal with the worst consequences of climate impacts which they have not caused while having few resources to do so. These two claims form the basis of the climate injustice argument.

A step toward climate justice was taken at long last at COP27 when the developing nations were able to successfully negotiate the creation of a climate justice fund targeted at loss and damage. Overcoming the opposition of the wealthy nations was a historic victory, but assuring the flow of money was pushed off to COP28 or beyond. Even with this step forward, COP27 overall did not tackle the root of the problem— that is, the burning of fossil fuel. A coalition of oil-producing nations, led by Saudi Arabia, were able to derail an attempt to have language included in the agreement that suggested the phaseout of oil and gas. The conference kept the language of commitment to warming of no more than 1.5° C, but without commitment to reduce the use of all fossil fuels obtaining that goal, lacking the massive introduction of carbon abating technology, remains elusive. Climate justice demands on behalf of the developing world, therefore, are likely to increase as more loss and damage accrues.

The climate justice demands of the youth, while heard by those negotiating world agreements, were not met. The demands by youth cannot be met, short of the world actually solving the climate crisis. After the first two decades of the 21st century, this remains an ongoing struggle.

CHAPTER 9

Emerging Climate Crisis Dimensions

Introduction

As the first quarter of the 21st century nears an end, the environment of climate crisis activism shows many signs of shifting emphasis while at the same time certain themes persist. U.S. climate activism in the first quarter of the 21st century moved from being a back burner issue to one that society recognized and voiced concern over. Climate denialism downscaled as more and more mainstream organizations and businesses accepted the reality of a future clean energy economy. That said, determining just how to achieve that vision continues to provoke widespread debate. This chapter will address that dialog both within the United States and internationally.

Beginning in 2020, the world experienced the rare event of a global pandemic. The spread of Covid-19 had huge disruptive effects that persisted for years. Economies shut down as nations tried to keep the virus from spreading. This massive disruption provided the opportunity for the climate crisis to be at the center of the pandemic recovery, and yet, it was not. The money that was poured into Covid recovery did not center on imaginative climate policies. There was, however, much to be learned from the pandemic that informs climate efforts. This chapter discusses these issues.

The first decades of the 21st century starkly revealed the cost of climate policy that resulted from the little progress made in lowering greenhouse gas (GHG) emissions. As the frequency and intensity of climate-related disasters mounted, these costs were repeatedly witnessed by populations worldwide. This chapter discusses these costs and the pressures they brought on both the private and public sectors to make the transition to a clean energy economy. Also, the signs of the energy transition underway are detailed.

Despite the progress made in the first quarter of the 21st century towards a worldwide effective climate policy, opposition was cemented. This chapter explores the persistence of the fossil fuel producers and others against the transition to a clean energy future and what their entrenchment means going forward.

Climate Activism Moves into the Mainstream

Zero Versus Net-Zero

U.S. climate activism in the year 2000 was decidedly less vocal and demanding than the activism that would appear a quarter of a century later. When Bill Clinton was president of the U.S., he knew that the treaty he had negotiated with other world leaders would be a nonstarter. Members of the Senate had voted unanimously to reject any treaty that might put the U.S. in a position to have to make disadvantageous economic decisions. Twenty-two years later, Joe Biden was able to push through a vote on what would be his signature climate effort, the Inflation Reduction Act. Climate policy had moved into the mainstream, at least within the Democratic party.

Part of the movement resulted from the increasing intense focus on events around the world—hurricanes, fires, droughts, heavy precipitation, flooding, heat waves, polar ice melts, and sea-level rise. Alarm about the climate crisis is a great motivator for action. One of the other reasons for the embracing of potential solutions by the mainstream was the emergence of the idea of net-zero emissions, as opposed to zero emissions. The difference between the two is stark. A zero emission policy would force deep decarbonization likely to be achieved only by stopping all fossil fuel use. Those that favor this approach tend to be activists that support rapid energy transition, many of whom fall far out of the political and economic mainstream. The other approach, net-zero, would support a transition to a lower-carbon energy system that would accommodate the net cancellation of emissions by mechanisms that absorb carbon while some emissions are allowed to continue (Yergin 2020).

The idea of net-zero emissions had always been a component in international negotiations of climate change, at least, of a sort. The Kyoto Protocol and the Paris Agreement included what negotiators called flexibility mechanisms that allowed the substitution of certain carbon dioxide (CO_2) emissions reductions projects undertaken in other countries for reductions at home. These mechanisms had generally been criticized by activists who argued that such projects, such as planting trees in developing nations, would occur without the incentives provided by the flexibility mechanisms. Critics also argued that the accounting for such projects was weak and that the reported reductions that states wanted credit for were not transparent or verifiable (Rahm 2019).

A shift occurred when the industry giants decided to be part of the net-zero commitment. For instance, the CEO of BP in February of 2020 announced that BP would be changing its role and pledged to reach net-zero emissions by 2050 (Kusnetz 2020). Other oil companies joined the effort including Shell and Enbridge (Berman and Taft 2021). By 2021, two-thirds of the world's biggest industrial emitters had signed on to net-zero pledges by 2050. These companies are collectively responsible for 80 percent of the world's emissions. Bloomberg estimates that these commitments, if achieved, would cut global greenhouse gas (GHG) emissions by the equivalent amount that China emits (Henze 2021). By January 2022, ExxonMobil joined the throng and declared it would have net-zero emissions by 2050 (Blackmon 2022).

While these announcements helped to drive the concept of net-zero mainstream, they were not without criticism. For instance, many oil companies claim that they can keep their current level of production or even expand production and achieve these commitments by planting trees and adopting largely unproven carbon-reduction technologies. The oil companies have mostly abandoned their former denial of the climate crisis, shifting to a posture that seeks to continue their business model while pledging net-zero emissions. Their net-zero narratives while mainstream, are unlikely to be achieved because many of the oil giants rely on what many consider unscientific analysis. The scientific analysis, activists insist, requires that to achieve maximum heating of 1.5° C, oil output must decline (Berman and Taft 2021). Moreover, the announcements by oil producers like ExxonMobil include a net-zero pledge that only covers their operations and not the main source of emissions from the use of oil and gas, which is the largest source of emissions from the industry (Thorbecke 2022).

Mass Demand for Climate Action

At the beginning of the century, climate activism was largely restricted to environmental and climate advocates like Al Gore, who since leaving office in 2001 began to focus his efforts on raising climate awareness. Gore established the Climate Reality Project in 2005 with the mission of engaging in grassroots leadership training to spread the word about climate change (Gore 2022). Other groups were also created to directly address climate issues, including scientific organizations. Early in the 21st century, traditional environmental organizations like Sierra Club, Friends of the Earth, WWF, and Greenpeace included climate change in their overall agendas as early as the 1990s when they became a party to the Climate Action Network (Climate Action Network 2022). However, by the second decade of the 21st century, the demand for climate action moved to the general public.

In the lead-up to the Copenhagen Summit in 2009, climate mobilization began to grow in scale. More than 40,000 people marched in Copenhagen demanding action on climate change (BBC News Channel 2009). On September 21, 2014, the People's Climate March was held in places around the world. The main target was New York City the day before the UN Climate Summit scheduled for the next day. It is estimated that hundreds of thousands of people marched in more than 150 cities worldwide. Secretary-General of the UN Ban Ki-moon joined the protest as did Jane Goodall and Al Gore (Davey, Vaughan and Holpunch 2014).

Again, before the COP21 meeting in Paris in 2015, the Global Climate March 2015 took place. More than 600,000 people from around the world took to the streets to demand that the talks in Paris be productive (Phipps, Vaughan and Milman 2015). In 2017, once again, the People's Climate March took place in Washington, D.C., to protest the Trump administration's climate policies. Hundreds of sister protests took place across the U.S. and around the world. The march took place on the 100th day of the Trump administration which had moved swiftly to roll back Obama's climate policy actions (Levenson 2017).

The other factor that drove the climate crisis into the mainstream was the relentless demand by youth and their supporters for climate action. Prior to the rise of the youth movement beginning in 2018 with Swedish activist Greta Thunberg's climate strike, climate protests were dominated by older people and scientists. Between 2018 and 2019 alone, Fridays for Future was responsible for organizing over 6,000 youth climate strikes in 185 countries (Rainsford and Saunders 2021). Climate activism greatly increased in 2019, and with it the attention of the public to climate matters. By analyzing internet search results in 45 countries between 2015 and 2019, researchers linked public interest in climate change to climate protests and activist activities (Sisco et al. 2021).

On September 20, 2019, the largest climate protest ever occurred around the world on the eve of the UN Climate Summit called by UN Secretary-General António Guterres to urge commitment to achieving the 1.5° C target set by the Paris Agreement. Young and old took to the streets in 185 countries to demand action. For the first time since the school strikes had begun in 2018, young people called adults to join them. The youth were joined by many older people and institutional groups. Trade unions mobilized support. Doctors and nurses joined the strike. Employees left their workplaces, including many from Google, Facebook, and Amazon. In all, more than one million people heard the call and came out (Laville and Watts 2019). The demand for climate action was broad-based and vocal.

Lessons from the Pandemic

When the pandemic drove countries into lockdown, the mass climate protests were affected. The youth, however, found new ways to spread their message. Abandoning the streets for massive Zoom and Twitter sessions, the youth adapted to social isolation and found new means of outreach (Morresi and Chulani 2021). Adults did the same.

Many raised the argument that the climate crisis should be at the center of the global recovery from the pandemic. As governments poured resources into the economic recovery, many suggested that such spending was an opportunity for visionary climate policies. Some world leaders heard the call. France's President Emmanuel Macron, for instance, refused to give stimulus funds to airlines that would not take steps to reduce GHG emissions. The argument was made that if governments were going to spend vast amounts of money to reboot their economies, it was only fair that they put those resources to rebuilding a greener, more sustainable, society. The idea was to align economic recovery with climate action (Espinosa 2020).

The world, however, missed this opportunity. Global GHG emissions from fossil fuels and industry did dip in 2020 but rose again by 2021. The size of the dip is important because it tells another lesson from the pandemic. In pre-pandemic 2019, global emissions were 37.08 billion tons. The dip during the lockdowns most countries experienced in 2020 dropped global emissions to 35.26 billion tons. But after nations began to unlock their economies, world emissions rose to 37.12 billion tons. If land use change is factored in, then emissions went from 41.64 billion tons in 2019, to 39.32 billion tons in 2020, to 41.06 billion tones in 2021 (Our World in Data 2022).

A significant lesson to be learned is that even with most of the world's economies in massive lockdown, emissions were barely affected. This fact raises awareness of the truly enormous economic changes that will have to occur for the world to get emissions down to where they need to be to create a sustainable planet. In terms of a permanent solution to climate change, these small declines in emissions during the lockdowns suggest that the only way to a permanent climate solution is to transition to a low-carbon economy so that economic activity can be maintained without GHG emissions (Rusmussen 2022).

But even the brief decline in emissions ushered in some welcome changes for the U.S. and other nations. Fewer cars on the roads resulted in dramatically better air quality for the U.S., UK, Italy, and China. Rivers and other waterways cleared bringing back fish to places like the Venice canals. Pollution in the Ganges declined by 30 percent. Dolphins were seen in the Bosphorus. With humans in retreat, more animals were seen returning to cities and towns. Cities became quieter. Fewer airplanes polluted the skies as air travel retracted. These results all suggest that if humans stop causing problems, the Earth can recover (In This Together 2020).

The pandemic also revealed the simple truth that if nations provide resources and lots of smart people go to work on the problem, solutions can quickly emerge. Of course, this is a lesson we did not need to endure the pandemic to learn, for we could turn to many historical instances of massive government investment in obtaining a desired result (landing on the moon, the Manhattan Project, etc.). For Covid, this was the creation of an effective vaccine in record time. One can only imagine the results if similar resources were leveraged to create the technologies that might effectively capture and sequester CO^2, or to hire an army of people to plant trees, or to engage in other behaviors scientists say will help.

Public reaction to the vaccine, especially in the U.S., is instructive. Even though the nation's public health system was fully supportive of the vaccines, misinformation and mistrust spread freely. Much of this was rooted in distrust of science itself, a phenomenon mirrored in the climate crisis debate. Vaccine hesitancy and denial of climate change share common themes: both are threats to personal, community, and global health; both require action rooted in social policy and cooperation; both rely on public trust of science. Getting those who reject science to change requires shifting human opinion patterns from doubt to belief and from belief to action. Education and improved public messaging are essential for this as is oversight over social media so dominant in spreading misinformation (Dobson 2022).

Related to this issue is the vaccine nationalism that emerged during the height of the pandemic, where wealthy nations denied the availability of a potential solution to poorer nations. This gap between rich and poor nations appears as well in the climate crisis whereby wealthy nations are hesitant to provide adequate finance and technology to the developing nations so that they might adapt and build resiliency. Moreover, just as Covid illuminated and further deepened the gap between rich and poor populations, so too does climate change.

Awareness of the Increasing Costs of the Climate Crisis

While the International Union of Geological Scientists still records the current era as the Holocene, many have labeled the 21st century the Anthropocene, that period of time when human activity fully dominates climate and the environment. The climate crisis looms large in this focus. Much of the scientific literature projects various scenarios that might play out. The "business as usual" or BAU scenario represents the worst case, where countries continue to burn fossil fuels relentlessly and temperatures rise by 4 or 5° C by the end of the century. The BAU scenario is compared with other possibilities, such as where greenhouse gas (GHG) emissions are diminished either greatly or by various amounts. These other scenarios have temperatures rising to a maximum of about 3° C. While there are scientific disagreements over the use of the BAU scenario, in particular the models including factors such as very high coal usage which is in decline worldwide, many scientists still think the BAU scenario might still be reached even with lower emissions if feedbacks in the climate system take effect. Climate feedbacks might occur when the Earth responds to warming by releasing more GHGs, for instance, processes such as the warming Arctic releasing more GHGs due to the thawing of the permafrost (Mooney and Freedman 2020).

Whatever the extent of the warming, climate-related damage will occur. Climate scientists historically have been hesitant to attribute any specific weather event to climate change, that is, until the rapid development of "attribution science" beginning in the 21st century. Attribution science allows scientists to provide quantitative estimates of how much the likelihood, frequency, or magnitude of an event was influenced by anthropogenic warming. Estimates can be provided quickly after an event (Hushaw 2017). As extreme weather events occur more frequently, the public asks the question—was this event caused by climate change?

Attribution science enables scientists to respond by saying how much climate change made an extreme event more severe, or more likely to occur, and by how much. Attribution science papers first began appearing in 2004 with a study of the European heat wave of 2003. The World

Weather Attribution (WWA) initiative subsequently developed. It is a collaboration of scientists who do analysis right after an extreme event occurs to determine how much climate change influenced the event. For example, in 2021, Belgium, the Netherlands, Germany, and Luxembourg experienced severe flooding that killed over 200 people. Historical studies of the region gave the event a 1 in 400 chance of occurring. Attribution studies done by WWA found that human-induced climate change made the event between 1.2 and 9 times more likely to occur than it would have 100 years ago. Warming also increased the rainfall amounts by 3 to 19 percent.

The heat wave that hit the Pacific northwestern U.S. in 2021 was analyzed. It was determined to be a one in a thousand years event that might never have occurred without climate change. The heat wave was also determined to be 2° C warmer than it would have been before the Industrial Revolution. If the world hits 2° C of warming, this type of one in a thousand years event could occur every 5 to 10 years. Hurricane Harvey was also studied, and researchers found that it was three times more likely to occur with anthropogenic climate change than without it, and that the rains were 15 percent greater with climate change than without it (Cho 2021).

Carbon Brief, a UK-based web site reporting recent developments in climate news, analyzed 350 peer-reviewed studies of extreme weather events worldwide. It found that extreme weather events increased in the decade beginning in 2010 and that 70 percent of 405 extreme weather events were made more likely or more extreme by climate change. Attribution science has increased the pressure on public officials in terms of the standard of care they provide and because it is allowing the quantification of risk, it will be used to assess damage in lawsuits. Attribution science enables the recognition of blame and the assignment of responsibility. It also provides information to communities and stakeholders so that they can take action to prevent harm (Cho 2021).

The U.S. alone has experienced $2.2 trillion in loses from hurricanes, severe storms, droughts, wildfires, and flooding since 1980. Climate change is making these events worse. Between 2017 and 2021, annual losses from billion dollar events totaled $765 billion, nearly eight times higher than in the 1980s. This number does not capture the cost of smaller less costly events and so does not show the true amount of economic distress caused. In 2021 alone, disaster was declared in the home county of 40 percent of Americans. Between 2017 and 2021, the U.S. experienced four of its most expensive wildfires, two of its three most expensive hurricanes, and its most severe winter storm (E2 2022).

A Pew Research Center survey of 10,282 American adults conducted May 2–8, 2022, showed that the public was well aware of the impacts of climate change. A majority of 71 percent of respondents said their community had experienced some form of extreme weather in the past year and that they see climate change as contributing to these events. For each of the extreme weather events included in the survey (heat, floods, drought, wildfires, and rising sea-levels) more than 80 percent of those who said they had been impacted blamed climate change for contributing to the event (Kennedy, Tyson and Funk 2022).

U.S. financial institutions are increasingly raising awareness of the risks of climate change. For instance, in November of 2020, the Federal Reserve identified climate change as a near-term financial instability risk. The Commodity Futures Trading Committee climate risk subcommittee issued a report calling climate change a major risk to the U.S. economy. The Securities and Exchange Commission named a senior official as their climate advisor and the commission called for public disclosure of climate vulnerability by U.S. firms. The Federal Housing Finance Agency's director made comments regarding climate change and the agency set up an Economics of Climate Summit to discuss the issue. The agency also hired key people to focus on the issue. The Treasury Secretary also made comments on climate change economic risks (CERES 2021).

Worldwide, weather-related disasters increased over the past 50 years. On average these losses amounted to 115 dead and $202 million USD in financial losses each day. Between 1970 and 2019, there were more than 11,000 reported disasters, with two million deaths and $3.64 trillion in loss. More than 91 percent of these deaths occurred in developing countries. Between 2010 and 2019 worldwide losses amounted to $383 million USD per day, a seven times increase from the 1970–1979 period. Looking forward, the World Meteorological Organization (WMO) Secretary-General stated that the number of weather, water, and heat extremes will worsen due to climate change. Attribution science studies of events between 2015 to 2017 show that 62 of 77 events show significant influence of human causation.

Regional analysis between 1970 to 2019 shows that Africa had nearly 1,700 extreme weather events in those 50 years that cost $38.5 billion. Asia experienced nearly 3,500 disasters costing $1.2 trillion. South America's top 10 events in the period resulted in economic damages of $39.2 billion. North America, Central America, and the Caribbean recorded nearly 2,000 disasters and economic damages of $1.7 trillion. The Southwest Pacific region experienced 1,400 disasters and economic damages of $163.7 billion. Europe experienced nearly 1,700 disasters, which cost a total of $476.5 billion (WMO 2021). The large scale of these disasters and economic damages has placed climate change and its impacts of the public agenda.

The role of developing nations in demanding climate reparations underscores the international understanding of the costs of climate impacts. At COP27 in 2022, developing countries led by Pakistan were able to push through their demand for loss and damage payments. This was seen as a major victory for developing countries who are experiencing the impacts of climate change in extreme ways and whose domestic budgets are insufficient to bear the burden (Ho and Farland 2022).

All these costs were repeatedly witnessed by populations worldwide. The suffering from human losses and economic damage could no longer be easily set aside. Awareness of the impacts of climate-related disasters moved from the small population of experts to the larger populations of citizens. A global awareness created an enhanced demand for action.

Signs of the Energy Transition Underway

Scholarly studies have detailed accounts of world energy transitions. The first of these was the shift from reliance on wood to coal in Elizabethan England and Ireland. The transition was forced, in large part, because the stock of wood was depleted. The transition to coal produced suffocating levels of urban air pollution but it also empowered a wave of innovation which led to the steam engine. By the late 1700s, the Industrial Revolution was well underway, powered initially by coal and the steam engine. At the same time, the adoption of gas and later oil for lighting, after the depletion of whale populations, resulted in the wider adoption of fossil fuels. Wood use peaked in the U.S. at 70 percent in 1870. It had peaked in Britain a century earlier. By 1900, coal use dominated 70 percent of U.S. demand and wood was in decline (Rhodes 2018). Worldwide, in 1900 coal supplied half of the world's energy demand.

Oil, discovered in Pennsylvania in 1859, took until the 1960s to supplant coal as the world's primary energy source. But even then, coal use continued to grow. The wide use of natural gas for power is more recent. History teaches that energy transitions occur, but progress can be slow in coming, and many innovations are needed to make the transitions a reality (Yergin 2020).

Moving the world to a low-carbon energy future will be one of the major challenges of the next decades. In 2020, 80 percent of worldwide fuel use was fossil fuels. Replacing that source of energy will require a transition to clean energy, including solar, wind, geothermal, hydropower, biomass, and nuclear. Not all of these sources of clean energy are universally acclaimed, for instance, nuclear energy is embraced by some and rejected by others due to concerns including safety and waste storage.

The adoption of new energy sources will have to be accompanied by new technologies, such as batteries and other forms of storage to make intermittent sources like solar and wind more useable for base energy supply. To address the emissions of GHGs more completely, improvements to energy efficiency will have to enhance structures, vehicles, appliances, industrial processes, and equipment. To achieve a net-zero low-carbon economy where some fossil fuel is still burned, technology will have to be developed and scaled for carbon capture and sequestration or storage (CCS) (EESI 2021).

The Growth of Clean Energy

Clean energy has seen remarkable growth in recent years. Led by solar and wind power, clean energy is projected to become the backbone of the world's power supply. Solar and wind started as industries dominated by small regional players whose costs were typically greater than coal power. However, new players have entered the sector including major oil and gas companies as well as large multinationals (including oil and gas companies) who seek to profit from the increasing demand for clean energy. Also entering the market are private-equity and institutional investors who are making solar and wind a major component of their strategy.

Additionally, costs for solar and wind have dropped significantly. Capacity additions in wind and solar have outperformed targets. Between 2017 and 2020, renewables have added 1.7 gigawatts of electricity annually. Estimates are that by 2026, global renewable-electric capacity will rise by more than 80 percent of 2020 levels, with two-thirds coming from solar and wind. Further estimates are that by 2035, renewables will generate 60 percent of the world's electricity.

But these advances will not come without challenges. First, there is a scarcity of top-quality land for development, especially for wind turbines and solar farms. Second, there is a shortage of blue- and white-collar employees. This is particularly a problem when projections are that by 2030 the solar and wind sector will have to hire 1.1 million blue-collar workers for construction and another 1.7 million for operation and maintenance. Third, supply chain problems threaten availability of steel and continued economic problems in China are causing manufacturing disruptions. Also, many of the raw materials needed for manufacture of wind and solar are predicted to be in short supply, including nickel, copper, and rare earths. Finally, some in the sector will be financially challenged by increasing competition while at the same time governments end their subsidies (Heinkek et al. 2022).

Global capacity increases will occur in all regions across the globe with China remaining the global leader in capacity additions. China is expected to reach 1200 GW of wind and solar in 2026—four years earlier than its target. India will double its rate of new installations compared to 2015–2020. Deployments in the U.S. and across Europe are expected to increase greatly by 2025. These four markets account for 80 percent of solar and wind growth worldwide. Solar photovoltaic (PV) remains the strongest area of growth. Onshore wind additions will be one quarter higher than in the 2015–2020 period and offshore wind capacity is set to triple by 2026 (IEA 2021). This growth sets the stage for the emergence of the new global energy economy.

The growth of clean energy will not be without turmoil. For instance, European wind industry manufacturers in 2022 faced huge losses because of Chinese competition, supply chain problems, and pricing setbacks. China's wind industry is growing and is vying for foreign markets, raising competition. China also manufactures many of the components used in turbines, giving them a hold on development worldwide. Moreover, European manufacturers face shakeups due to pricing strategies that have involved bidding low to get the project and then losing money on each delivery. The Inflation Reduction Act's incentives for domestic U.S. production also poses a threat to the European manufacturers as many look to relocate to the U.S. to take advantage of its provisions (Reed 2022). Turbulence is not uncommon in the development of new industries so much of this is to be expected, however, it adds to the complexity of development as new industries emerge and mature.

The Electric Vehicle (EV) Revolution

The electrification of road transport through the introduction of EVs will partner with the movement of the grid to renewables by phasing out another major emitter of GHGs, the internal combustion engine. While still a small part of all vehicles sales, in the second quarter of 2022, EV sales accounted for 5.6 percent of all sales in the U.S., up from 2.7 percent a year earlier. Government incentives through tax credits and deductions support the market while drivers are pleased with the clean energy, lower maintenance requirements, and improved performance. As the cost of gasoline soared in 2022 as a result of the Russia–Ukraine war (details below), buyers flocked to EVs. Businesses and municipalities are also looking to EVs as the fleet of the future. While limitations of driving range, recharging infrastructure, and recharging time continue to depress some sales, the movement to EVs is underway (Forbes 2022).

Government tax incentives assist in the market for EVs, but other government policies are equally important. In October of 2022, California banned the sale of new gas-powered vehicles by 2035. California has a history of causing changes in car manufacturing, simply because it is such a large market that manufacturers want to capture. For instance, in the 1990s, California demanded that a certain percentage of cars sold had to be zero-emission vehicles, thereby driving the industry to produce cleaner cars. This new rule will also push more and more auto manufacturers to move to electric vehicles. Other states may also adopt California's new standard as they did with California's zero-emission standard. Sales projections look promising, for instance, Ford's F150 Lightning truck by 2022 already had several hundred thousand preorders showing that the market demand is considerable. Nearly all automakers are currently selling or planning to sell EVs (Powell 2022).

The Russian Invasion of Ukraine and Its Impacts

In 2014, Moscow organized a referendum on "reunification" in Crimea, the Ukrainian peninsula with historical ties to Russia. Russia's President Vladimir Putin had sent in paramilitary forces to organize the effort that ended with the annexation of the territory. The entire country of Ukraine had been within the USSR during the Cold War. When the USSR dissolved, Ukraine became an independent state. While Ukraine returned to Russia the nuclear weapons in its territory, the pipeline infrastructure to transport gas remained in the country.

Putin, with the desire to return Russia to the status it held when it was the USSR, became concerned that Ukraine was moving away from Russia's influence and towards the West. This created tension between the westward-leaning Ukraine and Russia. As a result of the annexation of Crimea, the U.S. and the EU imposed sanctions on Russia. But in 2022, Russia launched a full attack on Ukraine, widening the war. That attack was followed by more sanctions and concern in Europe that Russian gas would cease to flow to it (Yergin 2020).

European dependence on Russia for natural gas had been an issue for some time. In 2015 the president of the European Union had argued that dependence on Russia's natural gas made Europe weak. In 2022, as part of its war efforts, Russia cut gas flows to Europe by nearly 90 percent forcing an energy crisis for Europe. Gazprom, the Russian-owned gas company, cut flows to Germany, blaming the problem on a leak which it said could not be fixed because of the many sanctions against Russia. The EU called the move energy blackmail that had resulted from the EU's continued support for Ukraine and sanctions against Russia. Europe, in response, lined up a host of alternative providers of supplies, including Liquid Natural Gas (LNG) coming from the U.S., as well as additional supplies from Norway and Azerbaijan. Germany also decided not to shut down some coal plants that it had planned to as part of its efforts to cut GHG emissions. Europe had managed to fill its storage capacity for the winter of 2022–23 but usage restrictions needed to be put into effect (Mettugh 2022).

Russia's gas shutoff may have proved successful in bringing in more money, however, it also provoked determination in Europe to end its dependence on Russian gas. In May of 2022, the European Commission announced a massive increase in solar and wind power and pledged to find an additional 210 billion Euros for investment over the next five years. While in the short term European response to the Russia-Ukraine war and Russia's imposed gas shortages will result in more coal burning, in the long run it will speed Europe's transition to clean energy. The EU proposed that 45 percent of the EU's energy mix should come from renewables by 2030, an increase over the earlier target of 40 percent. Russia's efforts, therefore, have resulted in a speedier European transition than previously planned (Ranklin 2022).

The Persistence of the Opposition to Delay or Obstruct Climate Action

Despite decades of increasingly clear evidence of the destructive effects of anthropogenic climate change, many actors still attempt to delay or obstruct any climate action. In the U.S. domestically, these include a wide array of special interests, fossil fuels and other large corporations, and the broad conservative countermovement. Internationally, the same groups are active, and they are joined by several oil-producing states seeking to assure a market for their products.

At the federal level in the U.S., barriers to climate action are raised by special interests including the oil, gas, and coal giants. For instance, during the 2019–2020 federal election cycle, the Koch Industries PAC,

Americans for Prosperity, spent more than $47.7 million in disclosed contributions. This does not include their "dark money" contributions which historically are much higher. The Koch brothers spent $80 billion on media campaigns aimed at debunking climate science. They were joined in these efforts by ExxonMobil. Large corporate interests also operate nationally in that they are major supporters of the conservative countermovement which has been a major obstructor of climate policies.

Beginning in the 1970s with climate science denial, the conservative countermovement grew. It is heavily supported by large corporate interests deriving profits from fossil fuels and other large corporate interests seeking to maintain the status quo. Its messages are frequently echoed by conservative news outlets, thus magnifying its voice (Basseches et al. 2022). Most of the groups that make up the conservative countermovement have abandoned their climate denialism; however, they have shifted to a tactic of obstruction and delay, efforts heavily funded by these special interests (Mann 2021).

A counterintuitive unhelpful force at the federal level has been the division within the Democratic coalition which on the whole supports climate action. When majorities are slim, one senator or a handful of representatives can command significant ability to control the agenda and form that any legislation might take. This was the case during Biden's first two years when Senator Joe Manchin (D-WV) was able to derail more progressive efforts to enact climate policy due to his allegiance to coal producers in his home state (Pierre and Neuman 2021).

At the state level in the U.S., obstacles include political party governance and institutional arrangements. Democratic governance tends to promote climate action while Republican governance obstructs it. Even if Democrats control the legislative branch, Republican governors can veto climate action. Institutional arrangements are also important. Institutional professionalism, which includes characteristics such as time in session, salary, and quality of staff, plays a role because climate change is a complex issue. The more professionalized the legislature, the more likely it will be able to act on climate issues.

More professionalized legislatures are also more adept at overcoming initiatives including model legislation from organizations like the American Legislative Exchange Council (ALEC), a business-affiliated conservative group active in writing anti-climate model legislation that can be adopted and advanced by elected legislators within each state.

States that have highly professionalized executive agencies with delegated regulatory power tend to be more effective in thwarting the lobbying efforts of moneyed special interests. A good example is

California's Air Resources Board. Another stumbling block can be the media within a state if they fail to raise climate change issues to a level of salience that drives policy makers to action. Interest groups at the state level can also serve as an impediment to policy action. Influential interest groups include fossil fuel and business lobbies, electric utilities, and the broad conservative countermovement.

States that are heavily dependent on fossil fuel companies for employment or tax revenue will tend to make concessions to them. Companies like ExxonMobil and Koch Industries have virtually unlimited money to spend on lobbying, media campaigns, and political contributions. At the state level, anti-climate trade group are active, including the American Petroleum Institute, the Oil Heat Institute, and chapters of the Chambers of Commerce (Basseches et al. 2022).

On the international level, obstruction comes from the same special interests active in the U.S. as well as from nations. The extent of special interest engagement with international talks can be seen by the attendance of more than 600 fossil fuel lobbyists at COP27 in 2022, up by 25 percent from COP26. The 636 registered fossil fuel lobbyists attending Sharm el-Sheikh composed a delegation larger than that of all but one country. These interests have great impacts on negotiations. At COP27 they were effective in slowing discussions on fossil fuel phase-out efforts.

A coalition of civil society groups submitted to the United Nations Framework Convention on Climate Change (UFCCC) a document arguing that discussions to meaningfully address climate change would fail as long as "polluting interests are granted unmitigated access to policy-making processes."

The United States Council on International Business pushed back on any discussion of limits to the number of corporate interests allowed in the talks, arguing it would "marginalize" a critical constituency (Michaelson 2022). Clearly, the strategy is to attempt to dominate negotiations with an eye to slowing processes and derailing any moves threatening to fossil fuel interests. Greta Thunberg voiced objection, saying, "If you want to address malaria, you don't invite the mosquitos" (Igini 2022).

Nations also play a role in obstruction. For instance, at COP27, a coalition of 80 developed and developing nations called for a phaseout of fossil fuels. India first proposed the language, and it was joined by the EU, UK, Australia, Canada, and the U.S. as well as by small island nations and some Latin American countries. But these efforts were thwarted by the oil-producing countries of Saudi Arabia, Iran, and Russia which wanted to continue to produce and sell their resources. This issue will persist and will likely be raised again at COP28 and later conferences (Ho and Farland 2022).

Conclusion

Over the next several decades certain emerging issues will dominate the climate crisis debate. The first of these is that the climate discussion has matured. It is no longer restricted to the realm of scientists and a small number of the attentive public. The climate crisis debate has gone mainstream and will continue to occupy a position of critical importance in international and national dialogues. This is so because the mass demand for action on climate change escalated enormously in the beginning of the 21st century. Youth and their allies played an important role in this intensification. The fact that politicians and businesses felt the pressure to pledge net-zero goals underscores the centrality of the issue. Activists may doubt that these net-zero targets are real and expect that they will have to push back hard on pledges that amount to little more than greenwashing.

While the climate crisis was a major issue in the first decades of the 21st century, it was not the only massive problem confronting humankind. The Covid-19 pandemic which blanketed the globe in 2020 raised issues of consequence for the climate crisis. The first of these was the transformation of networking over climate change to include internet-mediated approaches in addition to face-to-face activities. Activists learned new ways to organize and spread their messages. The pandemic also revealed the extent to which society must change to meet its goal of keeping warming to 1.5° C. In the pandemic, with most countries in some state of economic lockdown, emissions fell only slightly, and once lockdowns started to be lifted, emissions returned to the high levels of the pre-pandemic. The message was loud and clear for all listening.

Major shifts of enormous proportions will be needed to avoid the worst effects of climate change. The pandemic also exposed several other persistent problems—for instance, vaccine hesitancy tied to misinformation and failure to accept scientific advice. These attitudes easily relate to climate issues where misinformation and science denial are abundant. The pandemic also revealed something of national character when the wealthier nations adopted postures of vaccine nationalism, not unsimilar to the long continuing dispute between developed and developing nations over finance and technology.

Awareness of the increasing costs of the climate crisis grew substantially in the first decades of the 21st century as climate-related disasters multiplied. The pace of these disasters will likely continue as the century moves on, at least until the needed energy transition takes full form. The transition is underway with the amazing growth of the clean energy sector, particularly solar and wind, and with the electric vehicle transforming the automotive sector. Other technological developments to capture carbon are underway but need time to mature and reach a viable scale. The transition will not be as rapid as many hope. History shows that energy transitions, while quite doable, take time. There will be turmoil and setbacks along the way, as the Russian invasion of Ukraine and its impacts on energy show.

Climate action will not go forward without significant pressures to delay or obstruct it. Deeply entrenched forces act to impede shifting from behavior as usual. These include the oil and gas industries, other powerful companies that profit in the current environment, and political groups and movements. In the domestic U.S., these groups wield pressure not only at the federal level but also within the states, especially states dependent on fossil fuels for revenue and employment. Internationally, the same groups hold sway and are joined by nations dependent on their fossil fuel resources for economic growth and or viability. In all, this opposition is well organized, funded, and determined to slow or derail entirely progress toward a low-carbon world. The job of the activists will be a challenging one in the next decades, as they continue to press for their goals of a sustainable and livable planet for all.

References

Chapter 1

Arms Control Association. 2020. *Nuclear Weapons: Who Has What at a Glance.* August. Accessed January 6, 2021. https://www.armscontrol.org/factsheets/Nuclearweaponswhohaswhat.

Barnard, Anne. 2019. "A 'Climate Emergency' Was Declared in New York City. Will That Change Anything?" *New York Times*, July 5. Accessed November 1, 2020. https://www.nytimes.com/2019/07/05/nyregion/climate-emergency-nyc.html.

Baumgartner, Frank R., and Bryan D. Jones. 1993. *Agendas and Instability in American Politics.* Chicago: University of Chicago Press.

Baumgartner, Frank R., Suzanna L. De Boef, and Amber E. Boydstun. 2008. *The Decline of the Death Penalty and the Discovery of Innocence.* New York: Cambridge University Press.

Bloomberg, Michael, and Carl Pope. 2017. *Climate of Hope: How Cities, Businesses, and Citizens Can Save the Planet.* New York: St. Martin's Press.

Boydstun, A.E., and R.A. Glazier. n.d. "A Two-Tiered Method for Identifying Trends in Media Framing of Policy Issues: The Case of the War on Terror." *Policy Studies Journal* 41 (4): 707–736. https://doi.org/10.1111/psj.12038.

Britannica, the Editors of the Encyclopaedia. 2020. *Encyclopaedia Britannica.* February 5. Accessed January 6, 2021. https://www.nti.org/learn/treaties-and-regimes/treaties-between-united-states-america-and-union-soviet-socialist-republics-strategic-offensive-reductions-start-i-start-ii/.

Brugman, Britta C., and Christian Burgers. 2018. "Political Framing Across Disciplines: Evidence from 21st-Century Experiments." *Research and Politics* 1–7. doi:10.1177/2053168018783370.

Callaghan, Karen, and Frauke Schnell. 2005. "Introduction: Framing Political Issues in American Politics." In *Framing American Politics*, by Karen Callaghan and Frauke Schnell, 1–20. Pittsburgh: University of Pittsburgh Press.

Cann, Heather W., and Leigh Raymond. 2018. "Does Climate Denialism Still Matter? The Prevalence of Alternative Frames in Opposition to Climate Policy." *Environmental Politics* 27 (3): 433–458. doi:10.1080/09644016.2018.1439353.

Carrington, Damian. 2019. "Why the Guardian Is Changing the Language It Uses About the Environment." *The Guardian*, May 17. Accessed October 30, 2020. https://www.theguardian.com/environment/2019/may/17/why-the-guardian-is-changing-the-language-it-uses-about-the-environment.

Clark, Michael A., Nina G.G. Domingo, Kimberly Colgan, Sumil K. Thakrar, David Tilman, John Lynch, Inês Azevedo, and Jason D. Hill. 2020. "Global Food System Emissions Could Preclude Achieving the 1.5 and 2 degree C Climate Change Targets." *Science* 370 (6517): 705–708. doi:10.1126/science.aba7357.

climaterealityproject.org. 2019. "Why Do We Call It the Climate Crisis?" *The Climate Reality Project*, May 1. Accessed November 1, 2020. https://www.climaterealityproject.

org/blog/why-do-we-call-it-climate-crisis#:~:text=So%20when%20it%20comes%20 to,only%20term%20that%20makes%20sense.

Dewulf, Art. 2013. "Contrasting Frames in Policy Debates on Climate Change Adaptation." *Wires Climate Change* 4: 321–330. doi:10.1002/wcc.227.

Diamond, Emily P. 2020. "The Influence of Identity Salience on Framing Effectiveness: An Experiment." *Political Psychology* 41 (6): 1133–1150. doi:10.1111/pops.12669.

Editorial. 2019. "Recognising the Climate Crisis." *Hindustan Times*, November 24. Accessed November 1, 2020. https://www.hindustantimes.com/editorials/ recognising-the-climate-crisis-ht-editorial/story-v3GecFh9xed242exHphqhP.html.

Extinction Rebellion. n.d. Accessed November 2, 2020. https://rebellion.global/.

Friedman, Lisa. 2020a. "With John Kerry Pick, Biden Selects a 'Climate Envoy' with Stature." *The New York Times*, November 23. Accessed November 25, 2020. https:// www.nytimes.com/2020/11/23/climate/john-kerry-climate-change.html?action= click&module=Well&pgtype=Homepage&section=Climate%20and%20Environment.

_____. 2020b. "Biden Introduces His Climate Team." *The New York Times*, December 19. Accessed January 4, 2021. https://www.nytimes.com/2020/12/19/climate/biden-climate-team.html.

Gore, Al. 1992. *Earth in the Balance: Ecology and the Human Spirit.* Boston: Houghton Mifflin.

_____. 2006. *An Inconvenient Truth: The Planetary Emergency and What We Can Do About It.* Emmaus, PA: Rodale Press.

_____. 2007a. *An Inconvenient Truth: The Crisis of Global Warming.* New York: Viking.

_____. 2007b. "Nobel Lecture." Oslo, December 10. https://www.nobelprize.org/prizes/ peace/2007/gore/26118-al-gore-nobel-lecture-2007/.

Grandoni, Dino, and Alexandra Ellerbeck. 2020. "The Energy 202: Biden Sounds Signal He Is Serious About Climate Change with John Kerry Pick." *The Washington Post*, November 24. https://www.washingtonpost.com/politics/2020/11/24/energy-202-biden-sends-signal-he-is-serious-about-climate-change-with-john-kerry-pick/.

Green, Alison, and Molly Scott Cato. 2018. "Facts About Our Ecological Crisis Are Incontrovertible. We Must Take Action." *The Guardian*, October 26. Accessed November 2, 2020. https://www.theguardian.com/environment/2018/oct/26/facts-about-our-ecological-crisis-are-incontrovertible-we-must-take-action.

Hawken, Paul (Ed.). 2017. *Drawdown: The Most Comprehensive Plan Ever Proposed to Reversing Global Warming.* New York: Penguin.

Hodder, Patrick, and Brian Martin. 2009. "Climate Crisis? The Politics of Emergency Framing." *Economic & Political Weekly* XLIV (36): 53–60.

Horton, Joshua B. 2015. "The Emergency Framing of Solar Geoengineering: Time for a Different Approach." *The Anthropocene Review* 2 (2): 147–151. doi:10:1177/2053019 61557992.

Hulme, Michael. 2019. "Climate Emergency Politics Is Dangerous." *Issues in Science and Technology* 36 (1): 23–25. Accessed November 3, 2020. https://issues.org/ climate-emergency-politics-is-dangerous/.

Hurlbert, Margot, and Joyeeta Gupta. 2016. "Adaptive Governance, Uncertainty, and Risk: Policy Framing and Responses to Climate Change, Drought, and Flood." *Risk Analysis* 36 (2): 339–356. doi:10.1111/risa.12510.

Kelsey, Elin. 2020. *Hope Matters: Why Changing the Way We Think Is Critical to Solving the Environmental Crisis.* Vancouver: Greystone Books.

Kobayashi, Keiichi. 2020. "Emphasis Framing Effects of Conflicting Messages." *Journal of Media Psychology—Theories, Methods, and Applications* 32 (3): 119–129. Accessed December 31, 2020. doi:10.1027/1864-1105/a000263.

Lachapelle, Erick, Eric Montpetit, and Jean-Philippe Gauvin. 2014. "Public Perceptions of Expert Credibility on Policy Issues: The Role of Expert Framing and Political Worldviews." *Policy Studies Journal* 42 (4): 674–699.

Lee, Yu-Kang, and Chun-Tuan Chang. 2010. "Framing Public Policy: The Impacts of Political Sophistication and Nature of Public Policy." *The Social Science Journal* 47: 69–89.

Lenton, Timothy M., Johan Rockström, Owen Gaffney, Stefan Rahmstorf, Katherine Richardson, Will Steffen, and Hans Joachim Schellnhuber. 2019. "Climate Tipping Points—Too Risky to Bet Against." *Nature* 575 (7784): 592–595. doi:10.1038/d41586-019-03595-0.

Mann, Michael E. 2021. *The New Climate War: The Fight to Take Back Our Planet.* New York: Public Affairs/Perseus.

Marchese, David. 2020. "Greta Thunberg Hears Your Excuses. She Is Not Impressed." *The New York Times*, October 30. Accessed November 2, 2020. https://nytimes. com/interactive/2020/11/02/magazine/greta-thunberg-interview. html? action=click&module=Editors%20Picks&pgtype=Homepage.

Matthews, Mark K., Nick Bowlin, and Benjamin Hulac. 2018. "Inside the Sunrise Movement (It Didn't Happen by Accident)." *E&E News*, December 3. https://www.eenews. net/stories/1060108439.

McKibben, Bill. 1989. *The End of Nature.* New York: Random House.

Meyer, Robinson. 2018. "Democrats Establish a New House 'Climate Crisis' Committee." *The Atlantic*, December 28. Accessed October 30, 2020. https://www. theatlantic.com/science/archive/2018/12/house-democrats-form-new-committee-climate-crisis/579109/.

Morrison, Oliver. 2017. "The 'Most Famous Slideshow in the World'—Al Gore Presents His Dramatic PowerPoint on Climate Change in Pittsburgh." *PublicSource*, October 17. Accessed November 25, 2020. https://www.publicsource.org/the-most-famous-slideshow-in-the-world-al-gore-presents-his-dramatic-powerpoint-on-climate-change-in-pittsburgh/.

Mucciaroni, Gary, Kathleen Ferraiolo, and Megan E. Rubado. 2019. "Framing Morality Policy Issues: State Legislative Debates on Abortion Restrictions." *Policy Sciences* 52: 171–189. doi:10.1007/s1107-018-9336-2.

Nabi, Robin L., Nathan Walter, Neekaan Oshidary, Camille G. Endacott, Jessica Love-Nichols, Z.J. Lew, and Alex Aune. 2020. "Can Emotions Capture the Elusive Gain-Loss Framing Effect? A Meta-Analysis." *Communication Research* 47 (8): 1107–1130. doi:10.1177/0093650219861256.

Nelson, Thomas E., Rosalee A. Clawson, and Zoe Oxley. 1997. "Media Framing of a Civil Liberties Controversy and Its Effect on Tolerance." *American Political Science Review* 91 (3): 567–584.

Nie, Martin. 2003. "Drivers of Natural Resource-Based Political Conflict." *Policy Sciences* 36: 307–341.

NTI: Building a Safer World. October 26, 2011. Accessed January 6, 2021. https://www. nti.org/learn/treaties-and-regimes/treaties-between-united-states-america-and-union-soviet-socialist-republics-strategic-offensive-reductions-start-i-start-ii/.

Ploughshares Fund. 2020. Accessed January 6, 2021. https://www.ploughshares.org/topic/russia.

Rahm, Dianne. 2010. *Climate Change Policy in the United States: The Science, the Politics and the Prospects for Change.* Jefferson: McFarland.

_____. 2019. *U.S. Environmental Policy: Domestic and Global Perspectives.* St. Paul: West Academic Publishing.

Rahn, Wendy M., Sarah E. Gollust, and Xuyang Tang. 2017. "Framing Food Policy: The Case of Raw Milk." *Policy Studies Journal* 45 (2): 359–385.

Rigby, Sara. 2020. "Climate Change: Should We Change the Terminology?" *Science Focus*, February 3. Accessed November 1, 2020. https://www.sciencefocus.com/news/climate-change-should-we-change-the-terminology/.

Ripple, William J., Christopher Wolf, Thomas M. Newsome, Phoebe Barnard, and William R. Moomaw. 2020. "World Scientists' Warning of a Climate Emergency." *Bioscience* 70 (1): 8–12. doi:10.1093/biosci/biz008.

Risbey, James S. 2008. "The New Climate Discourse: Alarmist or Alarming?" *Global Environmental Change* 18: 26–37. http://sciencepolicy.colorado.edu/students/envs_4800/risbey_2008.pdf.

Sandoval, Michael. 2018. "Sunrise Movement Challenges House Democratic Leadership on 'Green New Deal.'" *The Sun Times*, November 21. Accessed November 2, 2020. https://www.fairfieldsuntimes.com/business/energy/sunrise-movement-challenges-house-democratic-leadership-on-green-new-deal/article_9c143ae4-edee-11e8-a6df-9f670dcbbfdd.html.

Schumacher-Matos, Edward. 2011. "Global Warming versus Climate Change: Does It Make a Difference?" *NPR Public Editor*, November 17. Accessed October 30, 2020. https://www.npr.org/sections/publiceditor/2011/11/17/142418671/global-warming-vs-climate-change-does-it-make-a-difference.

Sheffer, Lior, and Peter John Loewen. 2018. "Accountability, Framing Effects, and Risk-Seeking by Elected Representatives: An Experimental Study with American Local Politicians." *Political Research Quarterly* 72 (1): 49–62. doi:10.1177/10659 12918775252.

Sunrise Movement. n.d. Accessed November 2, 2020. https://www.sunrisemovement. org/about/?ms=AboutSunriseMovement.

Takach, Geo. 2019. "Climate Change or Climate Crisis? It's All in the Framing." *National Observer*, October 17. https://www.nationalobserver.com/2019/10/17/opinion/climate-change-or-climate-crisis-its-all-framing.

Tharoor, Ishaan. 2021. "Analysis: Biden Sweeps Away Trump's Climate-Change Denialism." *The Washington Post*, February 1. https://www.washingtonpost.com/world/2021/02/01/biden-climate-change-reversal/.

van Hulst, Merlijn, and Dvora Yanow. 2016. "From Policy 'Frames' to 'Framing': Theorizing a More Dynamic, Political Approach." *American Review of Public Administration* 46 (1): 92–112. doi:10.1177/0275074014533142.

Watts, Jonathan. 2018. "We Have 12 Years to Limit Climate Change Catastrophe, Warns UN." *The Guardian*, October 8. Accessed November 5, 2020. https://www.theguardian.com/environment/2018/oct/08/global-warming-must-not-exceed-15c-warns-landmark-un-report.

Wilson, Robert Evans. 2009. "The Most Powerful Motivator: How Fear Is Etched in Our Brains." September 23. Accessed November 6, 2020. https://www.psychologytoday.com/us/blog/the-main-ingredient/200909/the-most-powerful-motivator.

Yoder, Katte. 2019. "Is It Time to Retire 'Climate Change' for 'Climate Crisis'?" *Grist*, June 17. Accessed November 1, 2020. https://grist.org/article/is-it-time-to-retire-climate-change-for-climate-crisis/.

Zak, Dan. 2019. "How Should We Talk About What's Happening to Our Planet?" *The Washington Post*, August 27. Accessed October 30, 2020. https://www.washingtonpost.com/lifestyle/style/how-should-we-talk-about-whats-happening-to-our-planet/2019/08/26/d28c4bcc-b213-11e9-8f6c-7828e68cb15f_story.html.

Zhou, Naaman. 2019. "Oxford Dictionaries Declares 'Climate Emergency' the Word of 2019." *The Guardian*, November 21. Accessed November 1, 2020. https://www.theguardian.com/environment/2019/nov/21/oxford-dictionaries-declares-climate-emergency-the-word-of-2019.

Chapter 2

Abbott, Kenneth, and Steven Berstein. 2015. "The High-Level Political Forum on Sustainable Development: Orchestration by Default and Design." *Global Policy* 6 (3): 222–225.

Adam, Karla, and Rick Noack. 2021. "Young Climate Activists Join Greta Thunberg for First Major Fridays for Future Strikes of Pandemic." *The Washington Post*, September 9. Accessed September 9, 2021. https://www.washingtonpost.com/world/2021/09/24/fridays-future-greta-climate-protests/.

Adman, Per, and Katrin Uba. 2019. "Schoolchildren Around the World Are On Climate Strike. Here's What You Need to Know." *The Washington Post*, March 15. Accessed March 15, 2019. https://www.washingtonpost.com/politics/2019/03/15/schoolchildren-around-world-are-climate-strike-heres-what-you-need-know/.

Andresen, Steinar. 2007. "The Effectiveness of UN Environmental Institutions." *International Environmental Agreements: Politics, Law, and Economics* 7 (4): 317–336.

Audubon. 2022. *Audubon.* Accessed May 11, 2022. https://www.audubon.org/conservation/climate-initiative.

Biermann, Frank, and Bernd Siebenhener. 2009. "The Role and Relevance of International Bureaucracies." In *Managers of Global Change: The Influence of International Environmental Bureaucracies*, by Bernd Siebenhuner and Frank Biermann, 9–10. Cambridge: MIT Press.

Carrington, Damian. 2021. "'Blah, Blah, Blah': Greta Thunberg Lambasts Leaders Over Climate Crisis." *The Guardian*, September 28. Accessed September 28, 2021. https://www.theguardian.com/environment/2021/sep/28/blah-greta-thunberg-leaders-climate-crisis-co2-emissions.

Crow, Deserai A., and Andrea Lawlor. 2016. "Media in the Policy Process: Using Framing and Narratives to Understand Policy Influences." *Review of Policy Research* 33 (5): 472.

Dimock, Michael, and John Gramlich. 2021. *How America Changed During Donald Trump's Presidency.* Washington, DC: Pew Research Center.

Earth First! Journal. 2022. Accessed May 11, 2022. https://earthfirstjournal.news/about/.

Environmental Defense Fund. 2022. Accessed May 11, 2022. https://www.edf.org/.

Fauna and Flora International. 2022. May 23. Accessed May 23, 2022. https://www.fauna-flora.org/about/.

Fisher, Dana R. 2019. "The Youth Climate Summit Starts July 12. It's Full of Young Activists Trained in the Anti-Trump Movement." *The Washington Post*, July 12. https://www.washingtonpost.com/politics/2019/07/12/youth-climate-summit-starts-today-its-full-young-activists-trained-anti-trump-movement/.

Fridays for Future. 2018. Accessed May 18, 2022. https://fridaysforfuture.org/.

Fridays for Future U.S. 2022. Accessed May 18, 2022. https://fridaysforfutureusa.org/local-groups/.

Gambino, Lauren. 2019. "Greta Thunberg to Congress: 'You're Not Trying Hard Enough. Sorry.'" *The Guardian*, September 17. Accessed September 17, 2019. https://www.theguardian.com/environment/2019/sep/17/greta-thunberg-to-congress-youre-not-trying-hard-enough-sorry.

Goodell, Jeff. 2021. "What to Do About Jair Bolsonaro, the World's Most Dangerous Climate Denier." *Rolling Stone*, June 9. Accessed June 23, 2022. https://www.rollingstone.com/politics/politics-features/jair-bolsonaro-rainforest-destruction-1180129/.

Greenpeace. 2022. Accessed May 11, 2022. www.greenpeace.org.

Greenpeace International. 2022. May 20. Accessed May 20, 2022. https://www.greenpeace.org/international/.

Hays, Samuel P. 2000. *A History of Environmental Politics Since 1945.* Pittsburgh: University of Pittsburgh Press.

Hernandez, Emily. 2022. "Texas Warns Firms They Could Lose State Contracts for Divesting from Fossil Fuels." *The Texas Tribune*, March 16. Accessed June 22, 2022. https://www.texastribune.org/2022/03/16/texas-fossil-fuel-divestment-ban-inquiry/.

Hickman, Caroline, Elizabeth Marks, Panu Pihkala, Susan, Lewandoski, R. Eric Clayton, Elouise E. Mayall, Britt Wray, Catriona Mellor, and Lise von Susteren. 2021. "Climate Anxiety in Children and Young People and Their Beliefs About Government Responses to Climate Change: A Global Survey." *The Lancet*, December: 863–873.

International Union for the Conservation of Nature. May 23, 2022. Accessed May 23, 2022. https://www.iucn.org/about/iucn-a-brief-history.

Ivanova, Maria. 2010. "UNEP in Global Environmental Governance: Design, Leadership, Location." *Global Environmental Politics* 10 (1): 31–45.

———. 2013. "The Contested Legacy of Rio+20." *Global Environmental Politics* 13 (4): 4–6.

Lakhani, Nina. 2020. "'We Don't Have Any Choice': The Young Climate Activists Naming and Shaming Politicians." *The Guardian*, October 6. https://www.theguardian.com/environment/2020/oct/16/sunrise-movement-wide-awakes-us-climate-activists.

Lerer, Lisa, and Reid J. Epstein. 2021. "Abandon Trump? Deep in the G.O.P. Ranks, the MAGA Mind-Set Prevails." *New York Times*, January 21. Accessed May 10, 2022. https://www.nytimes.com/2021/01/14/us/politics/trump-republicans.html.

N.a. 2021. "Juliana v. United States: Ninth Circuit Holds That Developing and Supervising Plan to Mitigate Anthropogenic Climate Change Would Exceed Remedial Powers of Article III Court." *Harvard Law Review*. Accessed May 18, 2022. https://harvardlawreview.org/2021/03/juliana-v-united-states/.

Nargi, Lela. 2019. "Greta Thunberg's New York Visit Inspires Young Climate Activists." *The Washington Post*, September 9. Accessed September 9, 2019. https://www.washingtonpost.com/lifestyle/kidspost/greta-thunbergs-us-visit-inspires-young-climate-activists/2019/09/09/91a422f6-ce88-11e9-87fa-8501a456c003_story.html.

Nature Conservancy. 2022. Accessed May 11, 2022. https://www.edf.org/.

Nature Friends International. 2022. Accessed May 23, 2022. https://www.nf-int.org/en/about-us/mission-statement.

NRDC. 2022a. Accessed May 11, 2022. https://www.edf.org/.

_____. 2022b. Accessed May 11, 2022. https://www.worldwildlife.org/.

O'Neill, Kate. 2015. "Architects, Agitators, and Entrepreneurs." In *The Global Environment: Institutions, Law, and Policy*, by Regina S. Axelrod and Stacy D. VanDeveer, 27–28. Los Angeles: Sage.

Quackenbush, Casey. 2022. "The Climate Scientists Are Not Alright." *The Washington Post*, May 20. Accessed May 20, 2022. https://www.washingtonpost.com/climate-environment/2022/05/20/climate-change-scientists-protests/.

Rahm, Dianne. 2019. *U.S. Environmental Policy: Domestic and Global Perspectives*. St. Paul: West Academic Publishing.

Ricketts, Sam, Rita Cliffton, Loya Oduyeru, and Bill Holland. 2020. *States Are Laying a Road Map for Climate Leadership*. Washington, DC: Center for American Progress.

Rosenbaum, Water A. 2020. *Environmental Politics and Policy*. Los Angeles: Sage.

Roy, Diana. 2022. "Deforestation of Brazil's Amazon Has Reached a Record High. What's Being Done?" *Foreign Affairs*, March 17. Accessed June 23, 2022. https://www.cfr.org/in-brief/deforestation-brazils-amazon-has-reached-record-high-whats-being-done.

Schnaidt, Isabel, and Inma Galvez-Shorts. 2019. "Youthlead Blog Post." June 14. Accessed May 18, 2022. https://www.youthlead.org/resources/15-young-climate-activists-you-should-be-following-social-media.

"Senate Hearing 109–947." 2005. Accessed May 11, 2022. https://www.govinfo.gov/content/pkg/CHRG-109shrg32209/html/CHRG-109shrg32209.htm.

Sengupta, Somini. 2021. "Young Women Are Leading Climate Protests. Guess Who Runs Global Talks?" *The New York Times*, November 6. Accessed November 6, 2021. https://www.nytimes.com/2021/11/06/climate/climate-activists-glasgow-summit.html.

_____. 2022. "Four Take Aways on the Youth Climate Movement." *New York Times Climate Forward*, March 22. Accessed March 22, 2022. https://www.nytimes.com/2022/03/22/climate/youth-climate-protests.html.

Sierra Club. 2022. Accessed May 11, 2022. https://www.audubon.org/conservation/climate-initiative.

Smith, David. 2021. "'Tired of Broken Promises': Climate Activists Launch Hunger Strike Outside of White House." *The Guardian*, October 20. Accessed May 20, 2022. https://www.theguardian.com/environment/2021/oct/20/sunrise-movement-climate-activists-hunger-strike.

Sunrise Movement. n.d. Accessed May 18, 2022. https://www.sunrisemovement.org/.

Tharoor, Ishaan. 2019. "Bolsonaro, Trump and the Nationalists Ignoring Climate Disaster." *The Washington Post*, August 23. Accessed June 23, 2022. https://www.washingtonpost.com/world/2019/08/23/bolsonaro-trump-nationalists-ignoring-climate-disaster/.

350.org. 2009. Accessed May 18, 2022. https://350.org/about/.

Tindall, David, Mark C.J. Stoddart, and Riley E. Dunlap. 2021. "Climate Change Denial 2.0 Was on Full Display at COP26, But There Was Also Pushback." *The Conversation*, November 18. Accessed June 23, 2022. https://theconversation.com/climate-change-denial-2-0-was-on-full-display-at-cop26-but-there-was-also-pushback-171639.

United Nations Sustainable Development and Climate Agenda. 2022. Accessed May 25, 2022. https://www.un.org/en/our-work/support-sustainable-development-and-climate-action.

World Wildlife Fund. 2022. Accessed May 23, 2022. https://www.worldwildlife.org/about.

Yoder, Kate. 2021. "Too British for America? Extinction Rebellion Is Getting Lost in Translation." *Grist*, March 11. Accessed March 11, 2021. https://grist.org/climate/extinction-rebellion-growth-united-states/?utm_medium=email&utm_source=newletter&utm_campaign=daily.

Chapter 3

Al Gore. 2022. Accessed July 6, 2022. https://algore.com/about/the-climate-crisis.

Betsill, Michele M. 2001. "Mitigating Climate Change in US Cities: Opportunities and Obstacles." *Local Environment*, 393–406. Accessed August 1, 2022. doi:10.1080/13549830120091699.

Bodin, Madeline. 2019. "The Climate Fight Goes Local." *Planning*, April: 31–37.

Bradsher, Keith. 1999. "Ford Announces Its Withdrawal from Global Climate Coalition." *The New York Times*, December 8. Accessed August 3, 2022. https://www.nytimes.com/1999/12/07/business/ford-announces-its-withdrawal-from-global-climate-coalition.html.

Bush, George H.W. 1992. "Statement on Signing the Instrument of Ratification of the United Nations Framework Convention on Climate Change." *George H.W. Bush Presidential Library and Museum.* October 13. Accessed July 2, 2022. https://bush41library.tamu.edu/archives/public-papers/4953?fbclid=IwAR3vp0zzELT8zzmJL-RYqw6-qDY-h-c3o5D5Oo-vjpJ7M8Vkd9HfExUw6NE.

Caney, Simon. 2006. "Environmental Degradation, Reparations, and the Moral Significance of History." *Journal of Social Philosophy*, Fall: 464–482.

Carter, Jimmy. 2005. *Our Endangered Values: America's Moral Crisis.* New York: Simon & Schuster.

de Boer, Yvo. 2008. "Kyoto Protocol Reference Manual: On Accounting of Emissions and Assigned Amount." *UNFCCC*, November. Accessed July 6, 2022. https://unfccc.int/resource/docs/publications/08_unfccc_kp_ref_manual.pdf.

EPA. 2022. *Inventory of U.S. Greenhouse Emissions and Sinks.* April 14. Accessed July 2, 2022. https://www.epa.gov/ghgemissions/inventory-us-greenhouse-gas-emissions-and-sinks#:~:text=About%20the%20Emissions%20Inventory, Sinks%20since%20the%20early%201990s.

Gerwin, Virginia. 2002. "Climate Lobby Group Closes Down." *Nature*, February 7: 567.

Globalchange.gov. 2022. *About USGCRP.* Accessed July 2, 2022. https://www.globalchange.gov/about.

Hovi, Jon, Detlef F. Sprinz, and Guri Bang. 2010. "Why the United States Did Not Become a Party to the Kyoto Protocol: German, Norwegian, and US Perspectives." *European Journal of International Relations* 18 (1): 129–150. Accessed July 6, 2022. doi:https://doi.org/10.1177/1354066110380964.

IPCC. 2022a. *About the IPCC.* https://www.ipcc.ch/about/.

_____. 2022b. *History of the IPCC.* https://www.ipcc.ch/about/history/.

Lambright, W. Henry. 2008. "Government and Science: A Troubled, Critical Relationship and What Can Be Done About It." *Public Administration Review* 5–18.

McCright, Aaron M., and Riley E. Dunlap. 2000. "Challenging Global Warming as a Social Problem: An Analysis of the Conservative Movement's Counter-Claims." *Social Problems* 47 (4): 499–522. Accessed August 3, 2022. doi:10.2307/3097132.

McCright, Aaron M., and Riley E. Dunlap. 2003. "Defeating Kyoto: The Conservative Movement's Impact on US Climate Change Policy." *Social Problems* 50 (3): 348–373.

NASA. 2022. *Global Climate Change: Vital Signs of the Planet.* June 24. Accessed June 29, 2022. https://climate.nasa.gov/evidence/#:~:text=In%201896%2C%20a%20 seminal%20paper, Earth%27s%20atmosphere%20to%20global%20warming.

Rabe, Barry G. 2002. "Statehouse and Greenhouse: The States Are Taking the Lead on Climate Change." *Brookings*, March 1. Accessed July 26, 2022. https://www. brookings.edu/articles/statehouse-and-greenhouse-the-states-are-taking-the-lead-on-climate-change/.

Rahm, Dianne. 2010. *Climate Change Policy in the United States: The Science, the Politics and the Prospects for Change.* Jefferson: McFarland.

Rich, Nathaniel. 2019. *Losing Earth: A Recent History.* New York: Farrar, Straus and Giroux.

Sheppard, Kate. 2007. *Fifteen Green Religious Leaders.* Accessed August 2, 2022. https:// grist.org/article/religious/.

Shulman, Seth. 2007. *Smoke, Mirrors, and Hot Air: How ExxonMobil Uses Big Tobacco's Tactics to Manufacture Uncertainty on Climate Science.* Cambridge, MA: Union of Concerned Scientists.

Time. 1953. "Invisible Blanket." *Time,* May 25. Accessed July 26, 2022. https://web. archive.org/web/20090305040506/http://www.time.com/time/magazine/article/ 0,9171,890597,00.html?promoid=googlep.

UN. 1992. *United Nations Framework Convention on Climate Change.* https://unfccc. int/files/essential_background/background_publications_htmlpdf/application/pdf/ conveng.pdf.

_____. 1993. *Report of the United Nations Conference on Environment and Development.* Accessed June 30, 2022. https://documents-dds-ny.un.org/doc/UNDOC/GEN/ N92/836/55/PDF/N9283655.pdf?OpenElement.

_____. 2022a. *Conferences/Environment and Sustainable Development/Rio.* https:// www.un.org/en/conferences/environment/rio1992.

_____. 2022b. *Conferences/Environment and Sustainable Development/Stockholm.* https://www.un.org/en/conferences/environment/stockholm1972.

UNFCCC. 2020a. *Kyoto Protocol, Targets for the First Commitment Period.* Accessed July 6, 2022. https://unfccc.int/process-and-meetings/the-kyoto-protocol/what-is-the-kyoto-protocol/kyoto-protocol-targets-for-the-first-commitment-period.

_____. 2020b. *UNFCCC—25 Years of Effort and Achievement.* Accessed July 2, 2022. https://unfccc.int/timeline/.

Walker, Joe. 1998. "Draft Global Climate Science Communications Action Plan." Washington, DC: American Petroleum Institute.

Chapter 4

Ballotpedia. n.d. *Energy Policy Act of 2005.* Accessed August 16, 2022. https:// ballotpedia.org/Energy_Policy_Act_of_2005#:~:text=The%20Energy%20Policy%20 Act%20of%202005%20mandated%20that%20gasoline%20sold,billion%20 gallons%20of%20renewable%20fuels.

Bash, Dana, and Deirdre Walsh. 2006. "Bush Vetoes Embryonic Stem Cell Bill." September 25. Accessed August 18, 2022. https://www.cnn.com/2006/POLITICS/07/19/ stemcells.veto/.

Bohringer, Christoph, and Andreas Loschel. 2003. "Market Power and Hot Air in International Emissions Trading: The Impacts of the US Withdrawal from the Kyoto Protocol." *Applied Economics,* 651–663.

Boykoff, Maxwell T., and Jules M. Boykoff. 2007. "Climate Change and Journalistic Norms: A Case Study of US Media Coverage." *Geoforum* 38: 1190–1204.

Broad, William J. 2007. "From a Rapt Audience, a Call to Cool the Hype." *The New York Times,* March 13. Accessed August 20, 2022. https://www.nytimes.com/2007/03/13/ science/13gore.html.

Bush, George H.W. 1992. "Statement on Signing the Instrument of Ratification of the United Nations Framework Convention on Climate Change." *George H.W. Bush Presidential Library and Museum.* October 13. Accessed July 2, 2022. https://bush41library.tamu.edu/archives/public-papers/4953?fbclid=IwAR3vp0zz ELT8zzmJL-RYqw6-qDY-h-c3o5D5Oo-vjpJ7M8Vkd9HfExUw6NE.

Bush, George W. 2001. "President Bush's Letter to Three Senators Explaining His Rejection of the Kyoto Protocol." *Energy & Environment* 12 (4): 391–392. Accessed August 8, 2022. https://journals-sagepub-com.libproxy.txstate.edu/doi/pdf/10.1260/0958305011500760.

California Air Resources Board. 2019. *California & the Waiver: The Facts.* September 19. https://ww2.arb.ca.gov/resources/fact-sheets/california-waiver-facts.

Center for Climate and Energy Solutions. n.d. *Congress Climate History.* Accessed August 10, 2022. https://www.c2es.org/content/congress-climate-history/.

_____. 2022. *Regional Greenhouse Gas Initiative (RGGI).* Accessed August 20, 2022. https://www.c2es.org/content/regional-greenhouse-gas-initiative-rggi/.

Christie, Michael. 2001. "Outrage as US Dumps Kyoto." *The Sydney Daily Telegraph,* March 3: 27. Accessed August 10, 2022. https://bit.ly/3QyucjA.

Crabtree, Brad. 2008. Director, Great Plains Institute (March 7). Interview by Dianne Rahm.

de Boer, Yvo. 2008. "Kyoto Protocol Reference Manual: On Accounting of Emissions and Assigned Amount." *UNFCCC.* November. Accessed July 6, 2022. https://unfccc.int/resource/docs/publications/08_unfccc_kp_ref_manual.pdf.

DOE. 2007. *Midwestern Greenhouse Gas Reduction Accord.* Accessed August 22, 2022. https://www.osti.gov/biblio/21036864-midwestern-greenhouse-gas-reduction-accord#:~:text=The%20Midwestern%20Greenhouse%20Gas%20Reduction,emissions,and%20to%20combat%20climate%20change.

Dunlap, Riley E. 2008. *Climate-Change Views: Republican-Democratic Gaps Expand.* May 29. Accessed August 15, 2022. https://news.gallup.com/poll/107569/climate change-views-republicandemocratic-gaps-expand.aspx.

Egelko, Bob. 2007. "California's Emission-Control Law Upheld on 1st Test in US Court." *The San Francisco Chronicle*, December 13. Accessed August 18, 2022. https://www.sfgate.com/green/article/California-s-emission-control-law-upheld-on-1st-3300052.php.

EPA. 2022. *Inventory of U.S. Greenhouse Emissions and Sinks.* April 14. Accessed July 2, 2022. https://www.epa.gov/ghgemissions/inventory-us-greenhouse-gas-emissions-and-sinks#:~:text=About%20the%20Emissions%20Inventory, Sinks%20since%20 the%20early%201990s.

European Environmental Agency. 2019. *Energy Intensity.* December 5. Accessed August 15, 2022. https://www.eea.europa.eu/data-and-maps/indicators/total-primary-energy-intensity-3.

Fialka, John J. 2007. "States Want Higher Emissions Bar." *The Wall Street Journal*, April 4: A6.

Globalchange.gov. 2022. *About USGCRP.* Accessed July 2, 2022. https://www.global change.gov/about.

Greenhouse, Linda. 2007. "Justices Rule Against Bush Administration on Emissions." *The New York Times*, April 2. Accessed August 18, 2022. https://www.nytimes.com/2007/04/02/washington/02cnd-scotus.html.

Holtcamp, James A. 2007. "Dealing with Climate Change in the United States: The Non-Federal Response." *Journal of Land, Resources, and Environmental Law* 27: 79–86.

IEA. 2017. *Energy Efficiency and Security Act of 2007.* November 5. Accessed August 18, 2022. https://www.iea.org/policies/910-energy-independence-and-security-act-of-2007.

INFORSE. 2007. *IPCC 4th Report—Main Findings.* Accessed August 19, 2022. https://www.inforse.org/europe/dieret/Climate/IPCC%204th%20Report%20-%20main%20 findings.htm.

IPCC. 2022a. *About the IPCC*. https://www.ipcc.ch/about/.
_____. 2022b. *History of the IPCC*. https://www.ipcc.ch/about/history/.
Klare, Michael. 2004. "Bush-Cheney Energy Strategy: Procuring the Rest of the World's Oil." *Common Dreams News Center*, January 14. Accessed August 10, 2022. http://www.commondreams.org/views04/0113-01.htm.
Lambright, W. Henry. 2008. "Government and Science: A Troubled, Critical Relationship and What Can Be Done About It." *Public Administration Review*, January/February: 5–18.
Leslie, Mitch. 2004. "Sifting for Truth About Global Warming." *Science* 38: 2167–2167.
Luntz, Frank. 2003. "The Environment: A Cleaner, Safer, Healthier America." *Luntz Research Companies—Straight Talk*. Accessed August 19, 2022. https://www.sourcewatch.org/images/4/45/LuntzResearch.Memo.pdf.
Massachusetts v. Environmental Protection Agency. 2007. 549 US 1438.
NASA. 2022. *Global Climate Change: Vital Signs of the Planet*. June 24. Accessed June 29, 2022. https://climate.nasa.gov/evidence/#:~:text=In%201896%2C%20a%20seminal%20paper, Earth%27s%20atmosphere%20to%20global%20warming.
NOAA. 2022. *Climate Assessment for the Southwest (CLIMAS)*. Accessed August 22, 2022. https://climas.arizona.edu/research/southwest-climate-change-initiative-swcci.
NRDC. 2003. *West Coast Governors Launch New Global Warming Pollution Pact*. September 22. Accessed August 22, 2022. https://www.nrdc.org/media/2003/030922.
Ottinger, Richard L. 2010. "Copenhagen Climate Conference—Success or Failure." *Pace Environmental Law Review* 27 (2): 411–419.
Rahm, Dianne. 2010. *Climate Change Policy in the United States: The Science, the Politics and the Prospects for Change*. Jefferson: McFarland.
Reporters Committee for Freedom of the Press. 2005. *No Openness Required for Cheney Task Force*. May 12. Accessed August 10, 2022. https://www.rcfp.org/no-openness-required-cheney-energy-task-force/.
Revkin, Andrew C. 2006. "NASA's Goals Delete Mention of Home Planet." *The New York Times*, July 22. Accessed August 16, 2022. https://www.nytimes.com/2006/07/22/science/nasas-goals-delete-mention-of-home-planet.html.
_____. 2007a. "Memos Tell Officials How to Discuss Climate Change." *The New York Times*, March 8. Accessed August 16, 2022. https://www.nytimes.com/2007/03/08/washington/08polar.html.
_____. 2007b. "Climate Change Testimony Edited by White House." *The New York Times*, October 25. Accessed August 16, 2022. https://www.nytimes.com/2007/10/25/science/earth/25climate.html.
Rich, Nathaniel. 2019. *Losing Earth: A Recent History*. New York: Farrar, Straus and Giroux.
Roosevelt, Margot. 2008. "Lawsuit Targets EPA's Refusal." *Los Angeles Times*, January 3. Accessed August 18, 2022. https://www.latimes.com/archives/la-xpm-2008-jan-03-me-epa3-story.html.
Selin, Henrik, and Stacy D. Vandeveer. 2005. "Canadian-US Environmental Cooperation: Climate Change Networks and Regional Action." *American Review of Canadian Studies* 35(2): 353–378.
Shulman, Seth. 2007. *Smoke, Mirrors & Hot Air: How ExxonMobil Uses Big Tobacco's Tactics to Manufacture Uncertainty on Climate Science*. Cambridge, MA: Union of Concerned Scientists.
Sourcewatch. 2020. *Cheney Energy Task Force*. February 25. Accessed August 10, 2022. https://www.sourcewatch.org/index.php/Cheney_Energy_Task_Force.
Stolberg, Sheryl Gay. 2007. "Bush Vetoes Measure on Stem Cell Research." *The New York Times*, June 21. Accessed August 18, 2022. https://www.nytimes.com/2007/06/21/washington/21stem.html.
UN. 1992. *United Nations Framework Convention on Climate Change*. https://unfccc.int/files/essential_background/background_publications_htmlpdf/application/pdf/conveng.pdf.

_____. 1993. *Report of the United Nations Conference on Environment and Development.* Accessed June 30, 2022. https://documents-dds-ny.un.org/doc/UNDOC/GEN/N92/836/55/PDF/N9283655.pdf?OpenElement.

_____. 2022a. *Conferences/Environment and Sustainable Development/Rio.* https://www.un.org/en/conferences/environment/rio1992.

_____. 2022b. *Conferences/Environment and Sustainable Development/Stockholm.* https://www.un.org/en/conferences/environment/stockholm1972.

UNFCCC. 2020a. *Kyoto Protocol, Targets for the First Commitment Period.* Accessed July 6, 2022. https://unfccc.int/process-and-meetings/the-kyoto-protocol/what-is-the-kyoto-protocol/kyoto-protocol-targets-for-the-first-commitment-period.

_____. 2020b. *UNFCCC—25 Years of Effort and Achievement.* Accessed July 2, 2022. https://unfccc.int/timeline/.

United States Conference of Mayors. 2021. *Mayors Climate Protection Center.* Accessed August 22, 2022. https://www.usmayors.org/programs/mayors-climate-protection-center/.

University of Virginia Miller Center. 2022. *George W. Bush—Key Events.* Accessed August 10, 2022. https://millercenter.org/president/george-w-bush/key-events.

WCI. 2013. *Western Climate Initiative.* Accessed August 22, 2022. http://westernclimateinitiative.org/.

WECC. 2015. *Western Renewable Energy Generation Information System.* Accessed August 22, 2022. https://www.wecc.org/WREGIS/Pages/Default.aspx.

Weiner, Eric. 2007. "American Conscience Waking up to Climate Change." *National Public Radio*, December 6. Accessed August 20, 2022. https://www.npr.org/2007/07/07/11787222/american-conscience-waking-up-to-climate-change.

White House. 2008. *Energy Security for the 21st Century.* Accessed August 15, 2022. https://georgewbush-whitehouse.archives.gov/infocus/energy/#:~:text=In%202002%2C%20President%20Bush%20set,exploration%20for%20oil%20and%20gas.

Chapter 5

Broder, John M. 2009. "A Smaller, Faster Stimulus Plan, but Still With a Lot of Money." *The New York Times*, February 14. Accessed August 30, 2022. https://www.nytimes.com/2009/02/14/us/politics/14stimintro.ready.html.

Bromley-Trujello, Rebecca, and Mirya R. Holman. 2020. "Climate Change Policymaking in the States: A View at 2020." *Publius* 446–472.

Center for Climate and Energy Solutions. 2009a. *COP15 Copenhagen.* Accessed September 5, 2022. https://www.c2es.org/content/cop-15-copenhagen/.

_____. 2009b. *Waxman-Markey Short Summary.* June. Accessed August 30, 2022. https://www.c2es.org/document/waxman-markey-short-summary/.

_____. 2014. *COP 20 Lima.* Accessed September 5, 2022. https://www.c2es.org/content/cop-20-lima/.

_____. 2022. *Multi-state Initiatives.* Accessed September 15, 2022. https://www.c2es.org/content/multi-state-initiatives/.

Climate Policy Info Hub. 2015. *Observer NGOs and the International Negotiations.* February. Accessed September 15, 2022. https://climatepolicyinfohub.eu/observer-ngos-and-international-climate-negotiations.

CO2 Coalition. 2022. *About CO2 Coalition.* Accessed September 13, 2022. https://co2coalition.org/about/.

Davenport, Coral, and Eric Lipton. 2017. "How GOP Leaders Came to View Climate Change as Fake Science." *The New York Times*, June 3. Accessed September 13, 2022. https://www.nytimes.com/2017/06/03/us/politics/republican-leaders-climate-change.html.

DOE. 2022. *2009 American Recover and Reinvestment Act.* Accessed August 30, 2022. https://www.energy.gov/oe/information-center/recovery-act.

Echeverria, Daniella, and Philip Gass. 2014. "The United States and China's New

Climate Change Commitments: Elements, Implications and Reactions." *International Institute for Sustainable Development*, November: 1–4. Accessed September 2, 2022. https://www.iisd.org/system/files/publications/us-china-climate-change-commitments.pdf.

Ekanayake, Janada, Kithsiri Liyanage, Jianzhong Wu, Akihiko Yokoyama, and Nick Jenkins. 2012. *Smart Grid: Technology and Applications*. West Sussex, UK: John Wiley and Sons.

Federal Register. 2015. "Carbon Pollution Emission Guidelines for Existing Stationary Sources: Electric Utility Generating Units." *Federal Register*, October 23: 64662–64964. Accessed September 5, 2022. https://www.gpo.gov/fdsys/pkg/FR-2015-10-23/pdf/2015-22842.pdf.

Fiorino, Daniel J. 2022. "Climate Change and Right-Wing Populism in the US." *Environmental Politics*, 801–819. doi:10.1080/09644016.2021.2018854.

Goldenberg, Suzanne, and Dan Roberts. 2015. "Obama Rejects Keystone XL Pipeline and Hails US as Leader on Climate Change." *The Guardian*, November 6. Accessed September 1, 2022. https://obamawhitehouse.archives.gov/the-press-office/2013/06/25/remarks-president-climate-change.

Great Plains Institute. 2022. *Powering the Plains: Energy Transition Roadmap*. Accessed September 15, 2022. https://www.betterenergy.org/wp-content/uploads/2018/01/GPI_Powering_the_Plains.pdf.

Henry, Don. 2014. *A View from the 2014 UN Climate Summit in New York*. September 24. Accessed September 5, 2022. https://theconversation.com/a-view-from-the-2014-un-climate-summit-in-new-york-32118.

History. 2018. *Troubled Asset Relief Program*. February 1. Accessed September 19, 2022. https://www.history.com/topics/21st-century/troubled-asset-relief-program.

IPCC. 2013. *IPCC Reports*. Accessed September 6, 2022. https://www.ipcc.ch/reports/.

_____. 2014. *Summary for Policy Makers*. Accessed September 6, 2022. https://www.ipcc.ch/site/assets/uploads/2018/02/WG1AR5_SPM_FINAL.pdf.

James, Frank. 2009. *Climate Bill Passes House*. June 26. Accessed August 30, 2022. https://www.npr.org/sections/thetwo-way/2009/06/climate_bill_passes_in_the_hou.htm.

Kessler, Glenn. 2017. "When Did Mitch McConnell Say He Wanted to Make Obama a One-Term President?" *The Washington Post*, January 11. Accessed September 2, 2022. https://www.washingtonpost.com/news/fact-checker/wp/2017/01/11/when-did-mitch-mcconnell-say-he-wanted-to-make-obama-a-one-term-president/.

Library of Congress. 2015. *George C. Marshall Institute/CO2 Coalition*. Accessed September 13, 2022. https://www.loc.gov/item/lcwaN0002405/.

Lockwood, Matthew. 2018. "Right-Wing Populism and the Climate Change Agenda: Exploring the Linkages." *Environmental Politics* 27(4): 712–732.

Magill, Bobby. 2016. *The Suit Against the Clean Power Plan, Explained*. April 12. Accessed September 5, 2022. https://www.climatecentral.org/news/the-suit-against-the-clean-power-plan-explained-20234.

Obama, Barack. 2013. *Remarks by the President on Climate Change*. June 25. Accessed September 1, 2022. https://obamawhitehouse.archives.gov/the-press-office/2013/06/25/remarks-president-climate-change.

O'Harro, Robert, Jr. 2017. "A Two-Decade Crusade by Conservative Charities Fueled Trump's Exit from Paris Climate Accord." *The Guardian*, September 5. Accessed September 14, 2022. https://www.washingtonpost.com/investigations/a-two-decade-crusade-by-conservative-charities-fueled-trumps-exit-from-paris-climate-accord/2017/09/05/fcb8d9fe-6726-11e7-9928-22d00a47778f_story.html.

Pierre, Jeffrey, and Scott Neuman. 2021. "How Decades of Misinformation About Fossil Fuels Halted US Climate Policy." *NPR All Things Considered*, October 27. Accessed September 13, 2022. https://www.npr.org/2021/10/27/1047583610/once-again-the-u-s-has-failed-to-take-sweeping-climate-action-heres-why.

Rahm, Dianne. 2010. *Climate Change Policy in the United States: The Science, the Politics, and the Prospects for Change*. Jefferson: McFarland.

RGGI. 2022. *The Regional Greenhouse Gas Initiative.* Accessed September 15, 2022. https://www.rggi.org/sites/default/files/Uploads/Fact%20Sheets/RGGI_101_Factsheet.pdf.

The United States Conference of Mayors. 2022. *Mayors Climate Protection Center.* Accessed September 15, 2022. https://www.usmayors.org/programs/mayors-climate-protection-center/.

UNFCCC. 2015. *Adoption of the Paris Agreement.* December 12. Accessed September 5, 2022. https://unfccc.int/resource/docs/2015/cop21/eng/l09.pdf.

_____. 2016a. *Marrakech Action Proclamation for Our Climate and Sustainable Development.* November. Accessed September 5, 2022. https://unfccc.int/files/meetings/marrakech_nov_2016/application/pdf/marrakech_action_proclamation.pdf.

_____. 2016b. *Marrakech Partnership for Global Climate Action.* Accessed September 5, 2022. https://unfccc.int/files/paris_agreement/application/pdf/marrakech_partnership_for_global_climate_action.pdf.

Weiss, Daniel J. 2010. *Anatomy of a Senate Climate Bill Death.* October 12. Accessed August 30, 2022. https://www.americanprogress.org/article/anatomy-of-a-senate-climate-bill-death/.

White House of President Obama. 2014. *Year of Action.* December 31. Accessed September 2, 2022. https://obamawhitehouse.archives.gov/year-of-action.

Chapter 6

Aton, Adam. 2018. "Zinke Leaves Legacy of Weakened Environmental Protections." *Scientific American.* https://www.scientificamerican.com/article/zinke-leaves-legacy-of-weakened-environmental-protections/.

Ballotpedia. 2016. *Donald Trump Presidential Campaign, 2016 Energy and Environmental Policy.* Accessed September 23, 2022. https://ballotpedia.org/Donald_Trump_presidential_campaign,_2016_Energy_and_environmental_policy.

BBC News. 2020. "Climate Change: US Formally Withdraws From Paris Agreement." *BBC News,* November 4. Accessed October 3, 2022. https://www.bbc.com/news/science-environment-54797743.

Beauchamp, Zack. 2018. "Rex Tillerson Has Been Fired. Experts Say He Did Damage That Could Last 'A Generation.'" *Vox,* March 13. Accessed September 23, 2022. https://www.vox.com/world/2018/3/13/16029526/rex-tillerson-fired-state-department.

BLM. 2021. *Trump Administration Conducts First ANWR Coastal Plain Oil and Gas Lease Sale.* January 6. Accessed October 16, 2022. https://www.blm.gov/press-release/trump-administration-conducts-first-anwr-coastal-plain-oil-and-gas-lease-sale.

Bromley-Trujillo, Rebecca, and Mira R. Holman. 2020. "Climate Change Policymaking in the States: A View at 2020." *Publius,* May 7: 446–472.

Calmes, Jackie. 2021. "How the Republicans Have Packed the Courts for Years." *Time,* June 22. Accessed October 4, 2022. https://time.com/6074707/republicans-courts-congress-mcconnell/.

Center for Climate and Energy Solutions. 2022. *Multi-State Initiatives.* Accessed October 15, 2022. https://www.c2es.org/content/multi-state-initiatives/.

Cohen, Steve. 2022. "The Supreme Court and Radical Environmental Deregulation." *Columbia Climate School State of the Planet.* June 21. Accessed October 3, 2022. https://news.climate.columbia.edu/2022/06/21/the-supreme-court-and-radical-environmental-deregulation/.

Davenport, Coral. 2020a. "US to Announce Rollback of Auto Pollution Rules, a Key Effort to Fight Climate Change." *The New York Times,* March 30. Accessed October 5, 2022. https://www.nytimes.com/2020/03/30/climate/trump-fuel-economy.html.

_____. 2020b. "New Trump Rule Aims to Limit Tough Clean Air Measures Under Biden." *The New York Times,* December 9. Accessed October 5, 2022. https://www.nytimes.com/2020/12/09/climate/trump-pollution-regulations.html.

_____. 2020c. "Climate Change Legislation Included in Coronavirus Relief Deal." *The New York Times*, December 21. Accessed October 14, 2022. https://www.nytimes.com/2020/12/21/climate/climate-change-stimulus.html.

Davenport, Coral, Lisa Friedman, and Maggie Habberman. 2018. "EPA Chief Scott Pruitt Resigns Under Cloud of Ethics Scandals." *The New York Times*, July 5. Accessed September 27, 2022. https://www.nytimes.com/2018/07/05/climate/scott-pruitt-epa-trump.html.

Dillion, Jeremy, and Kelsey Brugger. 2019. "Rick Perry's Most Surprising Legacy as Energy Secretary Could Be Bigger Science Budget." *Science*. Accessed September 29, 2022. https://www.science.org/content/article/rick-perry-s-most-surprising-legacy-energy-secretary-could-be-bigger-science-budget.

EPA. 2022. *Vehicle Emissions California Waivers and Authorizations*. Accessed October 11, 2022. https://www.epa.gov/state-and-local-transportation/vehicle-emissions-california-waivers-and-authorizations.

Farah, Nina H. 2020. "'Energy Dominance' Under Fire as Court Revives Methane Rule." *GreenWire*, July 7. Accessed October 5, 2022. https://subscriber.politicopro.com/article/eenews/1063575103.

Finnegan, Conor, and Stephanie Ebbs. 2017. "A Tale of 2 US Delegations at Climate Talks." *ABC News*, November 14. Accessed October 17, 2022. https://abcnews.go.com/Politics/tale-us-delegations-climate-talks/story?id=51101699.

Fiorino, Daniel J. 2022. "Climate Change and Right-Wing Populism in the United States." *Environmental Politics* 31(5): 801–89. Accessed October 15, 2022. doi:10.1080/09644016.2021.2018854.

Forbes. 2017. *Profile: Rex Tillerson*. Accessed September 23, 2022. https://www.forbes.com/profile/rex-tillerson/?sh=181521ed704b.

Frazen, Rachel. 2020. "Trump Administration Resuming Coal Leasing on Public Lands." *The Hill*, February 26. Accessed October 6, 2022. https://thehill.com/policy/energy-environment/484854-trump-administration-resuming-coal-leasing-on-public-lands/.

Fridays for Future. 2022. *What We Do*. Accessed October 15, 2022. https://fridaysforfuture.org/.

Goldfuss, Christina. 2016. "Memorandum for Heads of Federal Departments and Agencies." *Council on Environmental Quality*, August 1. Accessed October 13, 2022. https://obamawhitehouse.archives.gov/sites/whitehouse.gov/files/documents/nepa_final_ghg_guidance.pdf.

Grandoni, Dino. 2020. "The Energy 202: How Amy Coney Barrett May Make It Harder for Environmentalists to Win in Court." *The Washington Post*, September 28. Accessed October 3, 2022. https://www.washingtonpost.com/politics/2020/09/28/energy-202-how-amy-coney-barrett-may-make-it-harder-environmentalists-win-court/.

Hardin, Sally. 2018. "Big Oil's Central Role in the Trump Administration's Culture of Corruption." *CAP Action*, August 28. Accessed October 15, 2022. https://www.americanprogressaction.org/article/big-oils-central-role-trump-administrations-culture-corruption/.

Institute for Policy Integrity. 2022. "Roundup: Trump-Era Agency Policy in the Courts." *New York School of Law*. April 25. Accessed October 4, 2022. https://policyintegrity.org/trump-court-roundup.

IPCC. 2018. "Global Warming of 1.5° C." https://www.ipcc.ch/sr15/chapter/spm/.

_____. 2021. "Summary for Policy Makers." Accessed October 14, 2022. https://www.ipcc.ch/report/ar6/wg1/downloads/report/IPCC_AR6_WGI_SPM.pdf.

_____. 2022. *Reports*. Accessed October 14, 2022. https://www.ipcc.ch/reports/.

Irfan, Umair. 2019. "Trump's EPA Just Replaced Obama's Signature Climate Policy with a Much Weaker Rule." *Vox*, June 19. Accessed October 3, 2022. https://www.vox.com/2019/6/19/18684054/climate-change-clean-power-plan-repeal-affordable-emissions.

Kaplan, Adiel. 2022. "Ryan Zinke Knowingly Misled Federal Investigators as Interior

Secretary, Inspector General Finds." *NBC News,* August 24. Accessed September 27, 2022. https://archive.epa.gov/epa/aboutepa/epa-administrator-wheeler.html.

Kelly, Caroline, and Michael Kosinski. 2019. "Pompeo Downplays Climate Change, Suggests 'People Move to Different Places.'" *CNN,* June 7. Accessed September 27, 2022. https://www.cnn.com/2019/06/07/politics/pompeo-climate-change-washington-times.

Kustin, Mary Ellen. 2019. "Bernhardt's Climate Track Record is Dangerous." *CAP,* March 25. Accessed September 29, 2022. https://www.americanprogress.org/article/bernhardts-climate-track-record-dangerous/.

Kuykendall, Taylor. 2020. "US Coal Jobs Down 24% for the Start of Trump Administration to Latest Quarter." *S&P Global,* November 20. Accessed October 15, 2022. https://www.spglobal.com/marketintelligence/en/news-insights/latest-news-headlines/us-coal-jobs-down-24-from-the-start-of-trump-administration-to-latest-quarter-61386963.

Matthews, Mark K., Bolin, Nick, and Benjamin Hulac. 2018. "Inside the Sunrise Movement (It Didn't Happen By Accident)." *E&E News,* December 3. Accessed October 15, 2022. https://subscriber.politicopro.com/article/eenews/1060108439.

Negin, Elliott. 2019. "10 Ways Andrew Wheeler Has Decimated EPA Protections in Just 1 Year." *The Equation,* July 15. Accessed September 27, 2022. https://archive.epa.gov/epa/aboutepa/epa-administrator-wheeler.html.

NPR. 2016. *CHARTS: Here's What Donald Trump Has Said on These Issues.* November 17. Accessed September 23, 2022. https://www.npr.org/2016/11/17/501582824/charts-heres-what-donald-trump-has-said-on-the-issues.

Plumer, Brad. 2018. "How Brett Kavanaugh Could Reshape Environmental Law From the Supreme Court." *The New York Times,* July 10. Accessed October 3, 2022. https://www.nytimes.com/2018/07/10/climate/kavanaugh-environment-supreme-court.html.

Popovich, Nadja, and Brad Plumer. 2020. "What Trump's Environmental Rollbacks Mean for Global Warming." *The New York Times,* September 17. Accessed October 5, 2022. https://www.nytimes.com/interactive/2020/09/17/climate/emissions-trump-rollbacks-deregulation.html.

Popovich, Nadja, Livia Albeck-Ripka, and Kendra Pierre-Louis. 2021. "The Trump Administration Rolled Back More Than 100 Environmental Rules. Here's the Full List." *The New York Times,* January 20. Accessed October 5, 2022. https://www.nytimes.com/interactive/2020/climate/trump-environment-rollbacks-list.html.

Rahm, Dianne. 2019. *US Environmental Policy: Domestic and Global Perspectives.* St. Paul: West Academic Publishing.

Reilly, Amanda. 2017. "Killing NEPA Guidelines Won't Stop Lawsuits over Impacts." *E&ENews,* March 29. Accessed October 6, 2022. https://subscriber.politicopro.com/article/eenews/1060052285.

Rennert, Kevin, Brian C. Prest, William A. Pizer, Richard G. Newell, David Anthoff, Cora Kingdon, Lisa Rennels, et al. 2021. "The Social Cost of Carbon: Advances in Long-Term Probabilistic Projections of Population, GDP, Emissions, and Discount Rates." *Brookings Papers on Economic Activity* 223–275. Accessed October 11, 2022. https://www.brookings.edu/wp-content/uploads/2021/09/15985-BPEA-BPEA-FA21_WEB_Rennert-et-al.pdf.

Reuters. 2017. "Trump Takes Aim at Obama Climate Action Plan—White House Website." *Reuters,* January 20. Accessed October 17, 2022. https://www.reuters.com/article/usa-trump-energy/trump-takes-aim-at-obama-climate-action-plan-white-house-website-idUSL1N1FA1BP.

Revkin, Andrew. 2017. "Trump's Attack on Social Cost of Carbon Could End Up Hurting Fossil Fuel Push." *Science,* August 24. Accessed October 5, 2022. https://www.science.org/content/article/trump-s-attack-social-cost-carbon-could-end-hurting-his-fossil-fuel-push.

Sabin Center for Climate Change Law. 2022. "President Trump Announces Withdrawal for Paris Agreement." *Columbia Law School.* Accessed October 3, 2022.

https://climate.law.columbia.edu/content/president-trump-announces-withdrawal-paris-agreement-0.

Schulman, Loren DeJonge. 2022. "Schedule F: An Unwelcome Resurgence." *Lawfare*, August 12. Accessed October 14, 2022. https://www.lawfareblog.com/schedule-f-unwelcome-resurgence#:~:text=The%20stated%20rationale%20for%20Schedule,impartiality%2C%20and%20sound%20judgment.%E2%80%9D.

Sidahmed, Mazin. 2016. "Climate Change Denial in the Trump Cabinet: Where do His Nominees Stand." *The Guardian*, December 15. Accessed September 29 2022. https://www.theguardian.com/environment/2016/dec/15/trump-cabinet-climate-change-deniers.

Silverman-Roati, Korey. 2021. "US Climate Litigation in the Age of Trump: Full Term." *Sabin Center for Climate Change Law*. June. Accessed October 14, 2022. https://climate.law.columbia.edu/sites/default/files/content/docs/Silverman-Roati%202021-06%20US%20Climate%20Litigation%20Trump%20Admin.pdf.

Sunrise Movement. 2022. *We Are The Climate Revolution*. Accessed October 15, 2022. https://www.sunrisemovement.org/.

Trump, Donald. 2022. "EO 13957 Creating Schedule F in the Excepted Service." *Federal Register*, October 16: 67631–67635. Accessed October 14, 2022. https://www.federalregister.gov/documents/2020/10/26/2020-23780/creating-schedule-f-in-the-excepted-service.

UVA/Miller Center. 2022. *Donald Trump—Key Events*. Accessed September 23, 2022. https://millercenter.org/president/trump/key-events.

Yang, Stephen. 2017. "Trump Signs Orders Advancing Keystone, Dakota Access Pipelines." *NBC News*, January 24. Accessed October 17, 2022. https://www.nbcnews.com/politics/white-house/trump-sign-orders-advancing-keystone-dakota-access-pipelines-n711321.

Yergin, Daniel. 2020. *The New Map: Energy, Climate, and the Clash of Nations*. New York: Penguin.

Chapter 7

Adam, Karle, Brady Dennis, and Annie Linskey. 2021. "As High-Stakes Climate Summit Begins, Biden Apologizes for US Withdrawal from Paris Accord." *The Washington Post*, November 1. Accessed November 2, 2022. https://www.washingtonpost.com/climate-environment/2021/11/01/cop26-glasgow-climate-live-updates/.

Ariza, Mario Alejandro, and Mose Buchele. 2022. "Texas Stumbles in Its Effort to Punish Green Financial Firms." *NPR*, April 29. Accessed November 4, 2022. https://www.npr.org/2022/04/29/1095137650/texas-stumbles-in-its-effort-to-punish-green-financial-firms#:~:text=Texas%20anti%2Ddivestment%20law%20against%20green%20investors%20stumbles%20%3A%20NPR&text=Texas%20anti%2Ddivestment%20law%20against%20green%20in.

Badlam, John, Stephen Clark, Suhrid Gajendragadkar, Adi Kumar, Sara O'Rourke, and Dale Swartz. 2022. "The CHIPS and Science Act: Here's What's in It." *McKinsey & Company*, October 4. Accessed November 2, 2022. CHIPS was authorized to spend $280 billion, the overwhelming bulk of which, $200 billion was earmarked for R&D and commercialization.

Bhatia, Aatish, Francesca Paris, and Margot Sanger-Katz. 2022. "See Everything the White House Wanted, and Everything It Got." *The New York Times*, October 20. Accessed November 4, 2022. https://www.nytimes.com/interactive/2022/10/20/upshot/biden-budget-before-after-animation.html.

Biden, Joseph. 2021a. "Executive Order on Protecting Health and the Environment and Restoring Science to Tackle the Climate Crisis." *The White House*, January 20. Accessed October 28, 2022. https://www.whitehouse.gov/briefing-room/presidential-actions/2021/01/20/executive-order-protecting-public-health-and-environment-and-restoring-science-to-tackle-climate-crisis/.

_____. 2021b. "Executive Order on Tackling the Climate Crisis At Home and Abroad." *The White House*, January 27. Accessed October 28, 2022. https://www.whitehouse. gov/briefing-room/presidential-actions/2021/01/27/executive-order-on-tackling-the-climate-crisis-at-home-and-abroad/.

Blinken, Antony J. 2021. "Remarks, UN Security Council Meeting on Climate and Security." *Department of State*, September 23. Accessed October 25, 2022. https://www. state.gov/secretary-antony-j-blinken-at-un-security-council-meeting-on-climate-and-security/.

BLS. 2022. *TED: The Economics Daily*, July 18. Accessed October 24, 2022. https://www. bls.gov/opub/ted/2022/consumer-prices-up-9-1-percent-over-the-year-ended-june-2022-largest-increase-in-40-years.htm.

Bureau of Labor Statistics. 2022. *Employment Situation—September 2022*. October 7. Accessed October 24, 2022. https://www.bls.gov/news.release/pdf/empsit.pdf.

C2ES. 2022. "Multi-State Initiatives." *C2ES*, November 7. Accessed November 7, 2022. https://www.c2es.org/content/multi-state-initiatives/.

Cattaneo, Lia. 2022. "WPA Revived Clean Car Waiver for California." *Environmental and Energy Law Program, Harvard University*, March 14. Accessed November 4, 2022. https://eelp.law.harvard.edu/2022/04/epas-revived-clean-cars-waiver-for-california/.

CDC. 2022. *CDC Museum: Covid- 19 Timeline*. Accessed October 24, 2022. https:// www.cdc.gov/museum/timeline/covid19.html#Mid-2020.

Cochrane, Emily, and Lisa Friedman. 2022. "What's in the Climate, Tax and Health Care Package." *The New York Times*, August 7. Accessed November 2, 2022. https:// www.nytimes.com/2022/08/07/us/politics/climate-tax-health-care-bill.html.

Committee for a Responsible Federal Budget. 2020. *Pete Buttigieg's Climate Change Plan*, February 28. Accessed October 24, 2022. https://www.crfb.org/blogs/pete-buttigiegs-climate-change-plan.

Curtis, Kevin. 2021. "How the American Rescue Plan Act Fights Climate Change." *NRDC*, March 21. Accessed November 2, 2022. https://www.nrdcactionfund. org/how-the-american-rescue-act-fights-climate-change/.

Davenport, Coral, Henry Fountain, and Lisa Friedman. 2021. "Biden Suspends Drilling Leases in Arctic National Wildlife Refuge." *The New York Times*, June 1. Accessed October 28, 2022. https://www.nytimes.com/2021/06/01/climate/biden-drilling-arctic-national-wildlife-refuge.html.

Denchak, Melissa, and Courtney Lindwall. 2022. "What Is the Keystone Pipeline?" *NRDC*, March 15. Accessed October 28, 2022. https://www.nrdc.org/stories/what-keystone-pipeline#:~:text=After%20more%20than%2010%20years,is%20now%20 gone%20for%20good.

Dennis, Brady, Chris Mooney, and Steven Mufson. 2022. "Climate Change Threatening 'Things Americans Value Most' US Report Says." *The Washington Post*, November 7. Accessed November 15, 2022. https://www.washingtonpost. com/climate-environment/2022/11/07/cop27-climate-change-report-us/.

Detrow, Scott, Tamara Keith, and Jennifer Ludden. 2020. "Biden to Name Gina McCarthy, Former EPA Chief, as Domestic Climate Coordinator." *NPR*, December 15. Accessed October 24, 2022. https://www.npr.org/sections/biden-transition-updates/2020/12/15/945937035/biden-to-name-gina-mccarthy-former-epa-chief-as-domestic-climate-coordinator.

DOI. 2021. "Secretary Deb Haaland." *Department of the Interior*. Accessed October 25, 2022. https://www.doi.gov/secretary-deb-haaland.

_____. 2022. *Biden-Harris Administration Announces First-Ever Offshore Wind Lease Sale in the Pacific*. October 18. Accessed November 1, 2022. https://www.doi.gov/ pressreleases/biden-harris-administration-announces-first-ever-offshore-wind-lease-sale-pacific.

Donovan, Doug. 2022. "US Officially Surpasses 1 Million Deaths." *Johns Hopkins Coronavirus Resource Center*, May 17. Accessed October 24, 2022. https://coronavirus. jhu.edu/from-our-experts/u-s-officially-surpasses-1-million-covid-19-deaths.

Drennen, Ari, and Sally Hardin. 2021. "Climate Deniers in the 117th Congress." *CAP*, March 30. Accessed November 7, 2022. https://www.americanprogress.org/article/climate-deniers-117th-congress/.

Enten, Harry. 2022. "How Joe Biden and the Democratic Party Defied Midterm History." *CNN*, November 13. Accessed November 28, 2022. https://www.cnn.com/2022/11/13/politics/democrats-biden-midterm-elections-senate-house.

EPA. 2021. "Michael S. Regan." *About EPA*. Accessed October 25, 2022. https://www.epa.gov/aboutepa/epa-administrator.

_____. 2022a. "Phasedown of Hydrofluorocarbons Final Rule Frequently Asked Questions." *EPA*, July 20. Accessed November 1, 2022. https://www.epa.gov/climate-hfcs-reduction/phasedown-hydrofluorocarbons-final-rule-frequently-asked-questions#:~:text=Consistent%20with%20the%20AIM%20Act,listed%20in%20the%20AIM%20Act.

_____. 2022b. *EPA Issues Supplemental Proposal to Reduce Methane and Other Harmful Pollution from Oil and Gas Operations.* November 11. Accessed November 14, 2022. https://www.epa.gov/controlling-air-pollution-oil-and-natural-gas-industry/epa-issues-supplemental-proposal-reduce#:~:text=November%2011%2C%202022%20%2D%2D%20EPA,existing%20oil%20and%20gas%20operations.

_____. 2022c. *Vehicle Emissions California Waivers and Authorizations.* Accessed November 4, 2022. https://www.epa.gov/state-and-local-transportation/vehicle-emissions-california-waivers-and-authorizations#:~:text=for%20that%20issue), Waiver%20Process, California's%20rules%20may%20be%20enforced.

Feng, Emily, John Ruwitch, and Franco Ordonez. 2022. "Biden and China's Xi Met for Three Hours, Here's What They Talked About." *NPR*, November 14. Accessed November 14, 2022. https://www.npr.org/2022/11/14/1136350988/biden-and-xi-are-meeting-in-bali-here-are-the-high-stakes-issues-on-the-agenda.

Flam, Faye. 2022. "The Omicron BA.5 Wave Is Starting to Ebb. We Need to Know Why." *The Washington Post*, August 11. Accessed October 24, 2022. https://www.washingtonpost.com/business/theomicronba5-wave-is-starting-to-ebb-we-need-to-know-why/2022/08/10/58b09a44-1910-11ed-b998-b2ab68f58468_story.html.

Friedman, Lisa. 2020. "With John Kerry Pick, Biden Selects a 'Climate Envoy' With Stature." *The New York Times*, November 23. Accessed October 24, 2022. https://www.nytimes.com/2020/11/23/climate/john-kerry-climate-change.html.

_____. 2022a. "Democrats Designed the Climate Law to Be a Game Changer. Here's How." *The New York Times*, August 22. Accessed November 4, 2022. https://www.nytimes.com/2022/08/22/climate/epa-supreme-court-pollution.html.

_____. 2022b. "Biden, Remaking Climate Team, Picks John Podesta to Guide Spending." *The New York Times*, September 2. Accessed September 2, 2022. https://www.nytimes.com/2022/09/02/climate/john-podesta-climate-biden.html.

Friedman, Lisa, and Jim Tankersley. 2022. "Biden Casts America as Climate Leader and Promises a 'Low-Carbon Future.'" *The New York Times*, November 11. Accessed November 14, 2022. https://www.nytimes.com/2022/11/11/climate/biden-cop27-climate-speech.html.

Gambino, Lauren. 2021. "'This Is Democracy Day': Biden Sworn in as 46th President of the United States." *The Guardian*, January 20. Accessed October 18, 2022. https://www.theguardian.com/us-news/2021/jan/20/joe-biden-sworn-in-46th-president-inauguration.

Gelles, David. 2022. "How Republicans Are 'Weaponizing' Public Office Against Climate Action." *The New York Times*, August 5. Accessed November 4, 2022. https://www.nytimes.com/2022/08/05/climate/republican-treasurers-climate-change.html.

Gillies, Rob. 2021. "Keystone XL Pipeline Halted as Biden Revokes Permit." *AP News*, January 20. Accessed October 28, 2022. https://apnews.com/article/joe-biden-alberta-2fbcce48372f5c29c3ae6f6f93907a6d.

Groom, Nichola, and Jennifer Hiller. 2021. "Biden Administration Pauses Federal Drilling Program in Climate Push." *Reuters*, January 21. Accessed October 28, 2022. https://

www.reuters.com/business/energy/biden-administration-pauses-federal-drilling-program-climate-push-2021-01-21/.

Harvey, Fiona. 2021. "Joe Biden Lambasts China for Xi's Absence from Climate Summit." *The Guardian*, November 2. Accessed November 2, 2022. https://www.theguardian.com/environment/2021/nov/02/cop26-joe-biden-lambasts-china-absence.

IPCC. 2022. *Reports*. September. Accessed November 7, 2022. https://www.americanprogress.org/article/climate-deniers-117th-congress/.

Joselow, Maxine. 2022. "State Climate Action Could be Supercharged by the Inflation Reduction Act." *The Washington Post*, August 17. Accessed November 2. https://www.washingtonpost.com/politics/2022/08/17/state-climate-action-could-be-supercharged-by-inflation-reduction-act/, 2022.

Karlamangla, Soumya. 2022. "What to Know About California's Ban on New Gasoline-Powered Cars." *The New York Times*, August 29. Accessed November 4, 2022. https://www.nytimes.com/2022/08/29/us/california-ban-gasoline-cars.html.

King, Ledyard. 2020. "'The Most Significant Climate Legislation Ever': How Stimulus Bill Tackles Warming Planet." *USA Today*, December 27. Accessed November 2, 2022. https://www.usatoday.com/story/news/politics/2020/12/27/covid-relief-legislation-includes-major-climate-change-provisions/4012433001/.

Kurtzleben, Danielle. 2021. "Ocasio-Cortez Sees Green New Deal Progress in Biden Plan, But 'It's Not Enough.'" *NPR*, April 2. Accessed November 7, 2022. https://www.npr.org/2021/04/02/983398361/green-new-deal-leaders-see-biden-climate-plans-as-a-victory-kind-of.

Liptak, Adam. 2022. "Supreme Court Limits EPA's Ability to Restrict Power Plant Emissions." *The New York Times*, June 30. Accessed November 3, 2022. https://www.nytimes.com/2022/06/30/us/epa-carbon-emissions-scotus.html.

Lobosco, Katie, and Tami Luhby. 2021. "Here's What's in the Bipartisan Infrastructure Package." *CNN*, November 15. Accessed November 2, 2022. https://www.cnn.com/2021/07/28/politics/infrastructure-bill-explained.

McGraath, Matt. 2021. "US Rejoins Paris Accord: Biden's First Act Sets Tone for Ambitious Approach." *BBC News*, February 19. Accessed October 28, 2022. https://www.bbc.com/news/science-environment-55732386.

Michigan.gov. 2007. "Granholm Says Addressing Climate Change Will Protect Environment, Create Jobs, Transform the State's Economy." *Michigan.gov*, December 12. Accessed October 25, 2022. https://www.michigan.gov/formergovernors/recent/granholm/press-releases/2007/12/12/says-addressing-climate-change-will-protect-environment-create-jobs-transform-states-economy.

Moore, Mark. 2021. "21 States Sue Biden for Rescinding Keystone XL Pipeline Permits." March 18. Accessed October 28, 2022. https://nypost.com/2021/03/18/21-states-sue-biden-over-keystone-xl-pipeline-permits/.

Mufson, Steven. 2022a. "The Surprising Political Shifts That Led to the Climate Bill's Passage." *The Washington Post*, August 13. Accessed November 2, 2022. https://www.washingtonpost.com/climate-environment/2022/08/13/surprising-political-shifts-led-climate-bills-passage/.

_____. 2022b. "US Ratifies Global Treaty Curbing Climate Super-Pollutants." *The Washington Post*, September 21. Accessed September 21, 2022. https://www.washingtonpost.com/climate-solutions/2022/09/21/kigali-amendment-senate-super-pollutants-climate/.

Newman, Rick. 2022. "Why the Inflation Reduction Act Is Called What It Is." *Yahoo! Finance*, August 15. Accessed November 4, 2022. https://finance.yahoo.com/news/why-the-inflation-reduction-act-is-called-what-it-is-200054655.html.

Phillips, Anna. 2022. "Federal Courts Reinstate Ban on New Coal Sales on Public Land." *The Washington Post*, August 12. Accessed November 4, 2022. ttps://www.washingtonpost.com/climate-environment/2022/08/12/court-coal-moratorium/.

Plumer, Brad, and Lisa Friedman. 2022. "Democrats Got a Climate Bill, Joe Manchin Got Drilling, and More." *The New York Times*, July 30. Accessed November 7, 2022. https://www.nytimes.com/2022/07/30/climate/manchin-climate-deal.html.

Rennert, Kevin, Brian C. Prest, William A. Pizer, Richard G. Newell, David Anthoff, Cora Kingdon, Lisa Rennels, et al. 2021. "The Social Cost of Carbon: Advances in Long-Term Probabilistic Projections of Population, GDP, Emissions, and Discount Rates." *Brookings Papers on Economic Activity* 223–275. Accessed November 4, 2022. https://www.brookings.edu/bpea-articles/the-social-cost-of-carbon/.

Reuters. 2022. "Biden Administration Finalizes Tougher Fuel Economy Rules." *CNBC*, April 1. Accessed November 4, 2022. ttps://www.washingtonpost.com/climate-environment/2022/08/12/court-coal-moratorium/.

RGGI. 2022. "Regional Greenhouse Gas Initiative." *C2ES*. November 7. Accessed November 7, 2022. https://www.c2es.org/content/regional-greenhouse-gas-initiative-rggi/.

Roberts, David. 2029. "The Green New Deal Explained." *Vox*, March 20. Accessed November 7, 2022. https://www.vox.com/energy-and-environment/2018/12/21/18144138/green-new-deal-alexandria-ocasio-cortez.

Sengupta, Somini, and Jenny Gross. 2022. "World Leaders Gather at COP27 Amid Compounding Crises of War, Warming and Economic Turmoil." *The New York Times*, November 7. Accessed November 7, 2022. https://www.nytimes.com/live/2022/11/07/climate/cop27-climate-summit.

Tabuchi, Hirolco, and Lisa Friedman. 2021. "Oil Executives Grilled Over Industry's Role in Climate Disinformation." *The New York Times*, October 28. Accessed November 4, 2022. https://www.nytimes.com/2021/10/28/climate/oil-executives-house-disinformation-testimony.html.

The White House. 2021a. "Justice 40." *The White House*. January. Accessed November 14, 2022. https://www.whitehouse.gov/environmentaljustice/justice40/.

_____. 2021b. "Press Briefing by Press Secretary Jen Psaki, Special Presidential Envoy for Climate John Kerry, and National Climate Advisor Gina McCarthy." *The White House*. January 27. Accessed October 28, 2022. https://www.whitehouse.gov/briefing-room/press-briefings/2021/01/27/press-briefing-by-press-secretary-jen-psaki-special-presidential-envoy-for-climate-john-kerry-and-national-climate-advisor-gina-mccarthy-january-27-2021/.

_____. 2021c. "Biden Invites 40 World Leaders to Leaders Summit on Climate." *The White House*, March 26. Accessed October 28, 2022. https://www.whitehouse.gov/briefing-room/statements-releases/2021/03/26/president-biden-invites-40-world-leaders-to-leaders-summit-on-climate/.

UN. 2022. *Climate Action*. November 7. Accessed November 7, 2022. https://www.un.org/en/climatechange.

US Department of State. 2021. "Leaders Summit on Climate." *US Department of State*. April 23. Accessed October 28, 2022. https://www.state.gov/leaders-summit-on-climate/.

US Department of Treasury. 2020. *About the CARES Act and the Consolidated Appropriations Act*. Accessed October 24, 2022. https://home.treasury.gov/policy-issues/coronavirus/about-the-cares-act.

Chapter 8

Adragna, Anthony. 2021. "Greta Thunberg to Testify in Congress on Earth Day." *Politico*, April 16. Accessed November 14, 2022. https://www.politico.com/news/2021/04/16/greta-thunberg-congress-testimony-482410.

BBC News. 2021. "COP26: What Was Agreed at the Glasgow Climate Conference." *BBC News*, November 15. Accessed November 17, 2022. https://www.bbc.com/news/science-environment-56901261.

Castle, Stephen. 2022. "As Climate Protests Get Bolder, British Police Strike Back with New Powers." *The New York Times*, November 20. Accessed November 21, 2022. https://www.nytimes.com/2022/11/20/world/europe/britain-protests-climate-change.html.

Climate Scorecard. 2021. "Climate Scorecard Country Ratings." *Climate Scorecard*.

Accessed November 17, 2022. https://www.climatescorecard.org/?gclid=Cj0KCQiA
1NebBhDDARIsAANiDD3Q8dJIS7691X_4krWTRaRuVNuosbiSMgktqdzwZwRAB
p5MRSi2AlAaAlhyEALw_wcB.

Evans, Simon, and Josh Gabbatiss. 2019. "COP25: Key Outcomes Agreed at the UN
Climate Talks in Madrid." *Carbon Brief.* December 15. Accessed November 18, 2022.
https://www.carbonbrief.org/cop25-key-outcomes-agreed-at-the-un-climate-talks-
in-madrid/.

Evans, Simon, and Jocelyn Timperley. 2018. "COP24: Key Outcomes Agreed at the
UN Climate Talks in Katowice." *Carbon Brief,* December 16. Accessed November 18,
2022. https://www.carbonbrief.org/cop24-key-outcomes-agreed-at-the-un-climate-
talks-in-katowice/.

Extinction Rebellion. 2022. "This Is an Emergency!" *Extinction Rebellion.* Accessed
November 11, 2022. https://rebellion.global/.

Fisher, Dana, and Sohana Nasrin. 2021. "Shifting Coalitions Within the Youth Climate
Movement in the US." *Politics and Government* 9 (2): 112–123. doi:10.17645/pag.
v9i2.3801.

Fridays for Future US. 2022. *Fridays for Future US.* Accessed November 11, 2022.
https://fridaysforfutureusa.org/.

Gonzalez, Carmen G., and Sumudu Atapatta. 2017. "International Environmental Law,
Environmental Justice, and the Global South." *Transnational Law & Contemporary
Problems* 26: 229–242.

Hayes, Christal. 2019. "'I Want You to Take Action': Greta Thunberg, Teenage Climate Activist, Testifies Before Congress." *USA Today*, September 18. https://www.
usatoday.com/story/news/politics/2019/09/18/greta-thunberg-climate-change-
testifies-congress/2357930001/.

Herr, Alexandria. 2020. "Meet UN Secretary General Antonio Guterres' New Youth
Climate Advisors." *Grist*, July 27. Accessed November 10, 2022. https://grist.org/
climate/meet-un-secretary-general-antonio-guterres-new-youth-climate-advisors/.

Hickman, Caroline, Elizabeth Marks, Panu Pihkala, Susan Clayton, R. Eric Lewand-
owski, Elouise E. Mayalt, Catriona Mellor, and Lise von Susteren. 2021. "Climate Anxiety in Children and Young People and Their Beliefs About Government
Responses to Climate Change: A Global Survey." *The Lancet* 5: 863–873.

Just Stop Oil. 2022. *Just Stop Oil.* Accessed November 15, 2022. https://juststopoil.org/.

Justice Democrats. 2022. *A Platform for Justice.* Accessed November 11, 2022. https://
justicedemocrats.com/platform/.

Kalmus, Peter. 2021. "Climate Depression Is Real. And It Is Spreading Fast Among
Our Youth." *The Guardian*, November 4. Accessed November 14, 2022. https://
www.theguardian.com/commentisfree/2021/nov/04/climate-depression-youth-
crisis-world-leaders.

Kaplan, Sarah. 2022. "COP27 Leaves the World on Dangerous Warming Path
Despite Historic Climate Fund." *The Washington Post*, November 20. Accessed
November 21, 2022. https://www.washingtonpost.com/climate-environment/2022/
11/20/cop27-climate-conference-deal-fund/.

New Consensus. 2022. *What Is New Consensus.* Accessed November 11, 2022. https://
newconsensus.com/about.

NPR. 2019. "Transcript: Great Thunberg's Speech at the UN Climate Action Summit." *NPR*, September 23. Accessed November 14, 2022. https://www.npr.
org/2019/09/23/763452863/transcript-greta-thunbergs-speech-at-the-u-n-
climate-action-summit#:~:text=%22You%20say%20you%20hear%20us,that%20
I%20refuse%20to%20believe.

Our Children's Trust. 2022a. "Legal Actions." *Our Children's Trust.* Accessed November 11, 2022. https://www.ourchildrenstrust.org/juliana-v-us.

_____. 2022b. "Our Children's Trust. Youth v. Gov." *Our Children's Trust.* Accessed
November 11, 2022. https://www.ourchildrenstrust.org/.

Rahm, Dianne. 2010. *Climate Change Policy in the United States: The Science, the Politics and the Prospects for Change.* Jefferson: McFarland.

_____. 2019. *US Environmental Policy: Domestic and Global Perspectives.* St. Paul: West Academic.

Rannard, Georgina. 2022. "COP27: Fresh Hope for Climate Talks After Climate Damage Offer." *BBC News*, November 17. Accessed November 17, 2022. https://www.bbc.com/news/science-environment-63677463.

Ritchie, Hannah. 2019. "Who Has Contributed Most to Global CO2 Emissions?" *Our World in Data.* October 1. Accessed November 16, 2022. https://ourworldindata.org/contributed-most-global-co2.

Ritchie, Hanna, and Max Roser. 2022. "Greenhouse Gas Emissions." *Our World in Data.* Accessed November 16, 2022. https://ourworldindata.org/grapher/annual-co2-emissions-per-country.

Rosenbaum, Walter A. 2020. *Environmental Politics and Policy.* Los Angeles: Sage.

Schools for Climate Action. 2022. *Climate Change Is a Generational Justice Issue.* Accessed November 11, 2022. https://schoolsforclimateaction.weebly.com/.

Schuppert, Fabian. 2012. "Climate Change & Intergenerational Justice." *UNICEF.* October 25. Accessed November 10, 2022. https://www.unicef-irc.org/article/920-climate-change-and-intergenerational-justice.html.

Shue, Henry. 2014. *Climate Justice: Vulnerability & Protection.* Oxford: Oxford University Press.

Stockholm Declaration. 1972. *Declaration of the United Nations Conference on the Human Environment.* Accessed November 15, 2022. file:///C:/Users/diann/Downloads/6471.pdf.

Sunrise Movement. 2022a. *About the Sunrise Movement.* Accessed November 11, 2022. https://www.sunrisemovement.org/about/.

_____. 2022b. *Fighting F*cking Fascism.* Accessed November 11, 2022. https://www.sunrisemovement.org/campaign/fcking-fighting-fascism/.

This Is Zero Hour. 2022. *We Can't Do This Alone.* Accessed November 11, 2022. https://www.thisiszerohour.org/platform.

350.org. 2022. "About 350." *350.org.* Accessed November 11, 2022. https://350.org/about/.

Thunberg, Greta. 2021. "Greta Thunberg Blah Blah Blah Speech, Milan 2021." *Carbon Independent.org*, September 28. Accessed November 14, 2022. https://www.carbonindependent.org/119.html.

United Nations. 1993. *Report of the United Nations Conference on Environment and Development.* New York: United Nations. https://documents-dds-ny.un.org/doc/UNDOC/GEN/N92/836/55/PDF/N9283655.pdf?OpenElement.

_____. 2017. "COP23: UN Climate Change Conference 2017." *United Nations.* Accessed November 17, 2022. https://www.un.org/sustainabledevelopment/cop23/.

_____. 2021. "The Glasgow Climate Pact—Key Outcomes from COP26." *United Nations.* December 13. Accessed November 17, 2022. https://unfccc.int/process-and-meetings/the-paris-agreement/the-glasgow-climate-pact-key-outcomes-from-cop26?gclid=Cj0KCQiA99ybBhD9ARIsALvZavWtO_IzIrahjqS_vkwwMITebpJ651Hugo EEzZDbjIh8Vg0NQ7f3dAYaAkzHEALw_wcB.

United Nations Climate Change. 2022a. *Action for Climate Empowerment.* November 10. Accessed November 10, 2022. https://unfccc.int/ace.

_____. 2022b. *Conference of Youth (COY).* November 10. Accessed November 10, 2022. https://unfccc.int/topics/education-and-youth/youngo/coy.

US Youth Climate Strike. 2020. "Action Network." *US Youth Climate Strike.* Accessed November 11, 2022. https://actionnetwork.org/groups/us-youth-climate-strike-3.

World Bank. 2021. *GDP Per Capita.* Accessed November 15, 2022. https://data.worldbank.org/indicator/NY.GDP.PCAP.CD?locations=CN.

Yeo, Sophie. 2016. "COP22: Key Outcomes Agreed at the UN Climate Talks in Marrakech." *Carbon Brief.* November 19. Accessed November 17, 2022. https://www.carbonbrief.org/cop22-key-outcomes-agreed-at-un-climate-talks-in-marrakech/.

Chapter 9

Basseches, Joshua A., Rebecca Bromley-Trujillo, Maxwell T. Boykoff, Trevor Culhane, Galen Hall, Noel Healy, David J. Hess, et al. 2022. "Climate Policy Conflict in the US States: A Critical Review and Way Forward." *Climate Change* 70 (32). Accessed November 26, 2022. doi:10.1007/s10584-022-03319-w.

BBC News Channel. 2009. *Climate Activists Condemn Copenhagen Police Tactics.* December 13. Accessed November 23, 2022. http://news.bbc.co.uk/1/hi/world/europe/8410414.stm.

Berman, Tzeporah, and Nathan Taft. 2021. "Global Oil Companies Have Committed to 'Net Zero' Emissions. It's a Sham." *The Guardian*, March 3. Accessed November 22, 2022. https://www.theguardian.com/commentisfree/2021/mar/03/global-oil-companies-have-committed-to-net-zero-emissions-its-a-sham.

Blackmon, David. 2022. "ExxonMobil Formally Joins the Net-Zero Bandwagon." *Forbes*, January 18. Accessed November 22, 2022. https://www.forbes.com/sites/davidblackmon/2022/01/18/exxonmobil-formally-joins-the-net-zero-by-2050-bandwagon/?sh=28d416250302.

CERES. 2021. "Turning Up the Heat: The Need for Urgent Actions by US Financial Regulators in Addressing Climate Risk." *CERES.* April 6. Accessed November 24, 2022. https://www.ceres.org/resources/reports/turning-heat-need-urgent-action-us-financial-regulators-addressing-climate-risk?utm_source=google-ad-grant&utm_medium=paid&utm_campaign=accelerator_finreg_keywords&utm_source_platform=googleadgrant&gclid=CjwKCAiAyfy.

Cho, Renee. 2021. "Attribution Science: Linking Climate Change to Extreme Weather." *Columbia Climate School.* October 4. Accessed November 24, 2022. https://news.climate.columbia.edu/2021/10/04/attribution-science-linking-climate-change-to-extreme-weather/.

Climate Action Network. 2022. *Tackling the Climate Crisis.* Accessed November 23, 2022. https://climatenetwork.org/.

Davey, Melissa, Adam Vaughan, and Amanda Holpunch. 2014. "People's Climate March: Thousands Demand Action Around the World—As It Happened." *The Guardian*, September 21. Accessed November 22, 2022. https://www.theguardian.com/environment/live/2014/sep/21/peoples-climate-march-live.

Dobson, Geoffrey P. 2022. "Wired to Doubt: Why People Fear Vaccines and Climate Change and Mistrust Science." *Frontiers in Medicine.* Accessed November 23, 2022. doi:10.3389/fmed.2021.809395.

EESI. 2021. "Fossil Fuels." *EESI*, July 22. Accessed November 25, 2022. https://www.eesi.org/topics/fossil-fuels/description#:~:text=Fossil%20fuels%E2%80%94including%20coal%2C%20oil,were%20compressed%20and%20heated%20underground.

Espinosa, Maria Fernanda. 2020. "The Climate Crisis Should Be at the Heart of the Global Covid Recovery." *The Guardian*, December 10. Accessed November 23, 2022. https://www.theguardian.com/commentisfree/2020/dec/10/the-climate-crisis-should-be-at-the-heart-of-the-global-covid-recovery.

E2. 2022. "Billion Dollar Losses, Trillion Dollar Threats: The Cost of Climate Change." *E2*, October 19. Accessed November 24, 2022. https://e2.org/reports/cost-of-climate-change/#:~:text=Billion%2Ddollar%20weather%20and%20climate,)%20(NOAA%2C%202022).

Forbes. 2022. "Growth Sector: Electric Vehicles Sales and the New Electric Economy Have Arrived." *Forbes*, September 24. Accessed November 25, 2022. https://www.forbes.com/sites/qai/2022/09/24/growth-sector-electric-vehicles-sales-and-the-new-electric-economy/?sh=5361964c143a.

Gore, Al. 2022. *Climate Reality Project.* Accessed November 23, 2022. https://algore.com/project/the-climate-reality-project.

Heinkek, Florian, Nadine Janecke, Holder Klarner, Florian Kuhn, Himayun Tai, and Raffael Winter. 2022. "Renewable-Energy Development in a Net-Zero World."

McKinsey & Company, October 28. Accessed November 25, 2022. https://www.mckinsey.com/industries/electric-power-and-natural-gas/our-insights/renewable-energy-development-in-a-net-zero-world.

Henze, Veronkia. 2021. "Two Thirds of the World's Heaviest Emitters Have Set a Net-Zero Target." *BloombergNEF*, September 24. Accessed November 22, 2022. https://about.bnef.com/blog/two-thirds-of-the-worlds-heaviest-emitters-have-set-a-net-zero-target/.

Ho, Joe, and Chloe Farland. 2022. "Late Night Fossil Fuel Fight Leaves Bitter Taste After COP27." *Climate Home News*, November 24. Accessed November 26, 2022. https://www.climatechangenews.com/2022/11/24/late-night-fossil-fuel-fight-leaves-bitter-taste-after-cop27/.

Hushaw, Jennifer. 2017. "Attributing Extreme Events to Climate Change." *Manomet*, February 16. Accessed November 24, 2022. https://www.manomet.org/publication/attributing-extreme-events-to-climate-change/?gclid=CjwKCAiAyfybBhBKEiwAgtB7fk4QBLSKcQqNW0EkWJ3nyLllzShwbHRG0JAAzoGkAyLO6RILVAsoIhoCdNYQAvD_BwE.

IEA. 2021. "Renewable Electricity Growth Is Accelerating Faster Than Ever Worldwide, Supporting the Emergence of the New Global Energy Economy." *IEA*, December 1. Accessed November 25, 2022. https://www.iea.org/news/renewable-electricity-growth-is-accelerating-faster-than-ever-worldwide-supporting-the-emergence-of-the-new-global-energy-economy.

Igini, Martina. 2022. "More Than 600 Fossil Fuel Lobbyists Join COP27 While Reports Predict Record Emissions in 2022." *Earth.org*, November 11. Accessed November 26, 2022. https://earth.org/fossil-fuel-lobbyists-cop27/#:~:text=Swedish%20climate%20activist%20Greta%20Thunberg,ways%20to%20halt%20global%20warming.

In This Together. 2020. "5 Ways Earth Started Healing During the Lockdown." *In This Together*, July 29. Accessed November 23, 2022. https://inthistogetheramerica.org/2020/07/29/5-ways-earth-started-healing-during-lockdown/?gclid=CjwKCAiApvebBhAvEiwAe7mHSLPtGqlLAtZQv7oZ9XMIqOXMVm98QUGL095_yMLiMaLWLGrj4bmlkhoCQlMQAvD_BwE.

Kennedy, Brian, Alec Tyson, and Cary Funk. 2022. *Americans Divided Over Direction of Biden's Climate Change Policies*. Washington, DC: Pew Research Center, July 14. Accessed November 26, 2022. https://www.pewresearch.org/science/2022/07/14/americans-divided-over-direction-of-bidens-climate-change-policies/.

Kusnetz, Nicholas. 2020. "What Does Net Zero Emissions Mean for Big Oil? Not What You'd Think." *Inside Climate News*, July 16. Accessed November 22, 2022. https://insideclimatenews.org/news/16072020/oil-gas-climate-pledges-bp-shell-exxon/?gclid=Cj0KCQiAg_KbBhDLARIsANx7wAzAILCf9l_1HT68LMtDOZ5uJqHQ-ZbkMNV3Glqtwh-HE03TAfoI3EgaAgy2EALw_wcB.

Laville, Sandra, and Jonathon Watts. 2019. "Across the Globe, Millions Join Biggest Climate Protest Ever." *The Guardian*, September 20. Accessed November 23, 2022. https://www.theguardian.com/environment/2019/sep/21/across-the-globe-millions-join-biggest-climate-protest-ever.

Levenson, Eric. 2017. "Climate Protest Take on Trump's Policies—and the Heat—In DC March." *CNN*, April 29. Accessed November 22, 2022. https://www.cnn.com/2017/04/29/us/climate-change-march.

Mann, Michael E. 2021. *The New Climate Wars: The Fight to Take Back Our Planet*. New York: PublicAffairs.

Mettugh, David. 2022. "Explainer: Europe Struggles to Cope with Russia Gas Shutoffs." *AP*, September 9. Accessed November 25, 2022. https://apnews.com/article/russia-ukraine-france-germany-prices-da1d935fa8bcba4c283f7c5b559a5c9a.

Michaelson, Ruth. 2022. "'Explosion' in Number of Fossil Fuel Lobbyists at COP27 Climate Summit." *The Guardian*, November 10. Accessed November 26, 2022. https://www.theguardian.com/environment/2022/nov/10/big-rise-in-number-of-fossil-fuel-lobbyists-at-cop27-climate-summit.

Mooney, Chris, and Andrew Freedman. 2020. "We May Avoid the Very Worst Climate

Scenario. But the Next-Worse Is Still Pretty Awful." *The Washington Post*, January 30. Accessed November 24, 2022. https://www.washingtonpost.com/weather/2020/01/30/we-may-avoid-very-worst-climate-scenario-next-worst-is-still-pretty-awful/.

Morresi, Elena, and Nikhita Chulani. 2021. "Protest in a Pandemic: Voices of Young Activists—Video." *The Guardian*, April 23. https://www.theguardian.com/environment/video/2021/apr/23/voices-of-young-climate-activists-how-the-pandemic-changed-the-way-we-protest-video.

Our World in Data. 2022. "CO2 Emissions." *Our World in Data*. Accessed November 23, 2022. https://ourworldindata.org/co2-emissions.

Phipps, Claire, Adam Vaughan, and Oliver Milman. 2015. "Global Climate March 2015: Hundreds of Thousands March Around the World—As It Happened." *The Guardian*, November 29. Accessed November 22, 2022. https://www.theguardian.com/environment/live/2015/nov/29/global-peoples-climate-change-march-2015-day-of-action-live.

Pierre, Jeffrey, and Scott Neuman. 2021. "How Decades of Disinformation About Fossil Fuels Halted US Climate Policy." *NPR*, October 27. Accessed November 26, 2022. https://www.npr.org/2021/10/27/1047583610/once-again-the-u-s-has-failed-to-take-sweeping-climate-action-heres-why.

Powell, Alvin. 2022. "California Dreaming? Nope." *The Harvard Gazette*, September 9. Accessed November 25, 2022. https://news.harvard.edu/gazette/story/2022/09/what-to-expect-from-california-gas-powered-car-ban/#:~:text=Last%20month%2C%20California%20regulators%20passed,the%20fight%20against%20climate%20change.

Rahm, Dianne. 2019. *US Environmental Policy: Domestic and Global Perspectives.* St. Paul: West Academic.

Rainsford, Emily, and Clare Saunders. 2021. "Young Climate Protesters' Mobilization Availability: Climate Marches and School Strikes Compared." *Frontiers in Political Science*. Accessed November 22, 2022. doi:10.3389/fpos.2021.71334.

Ranklin, Jennifer. 2022. "EU Plans 'Massive' Increase in Green Energy to Help End Reliance on Russia." *The Guardian*, May 18. Accessed November 25, 2022. https://www.theguardian.com/environment/2022/may/18/eu-plans-massive-increase-in-green-energy-to-rid-itself-of-reliance-on-russia.

Reed, Stanley. 2022. "Europe's Wind Industry Is Stumbling When It's Needed Most." *The New York Times*, November 22. Accessed November 26, 2022. https://www.nytimes.com/2022/11/22/business/wind-power-europe.html.

Rhodes, Richard. 2018. *Energy: A Human History.* New York: Simon & Schuster.

Rusmussen, Carol. 2022. "A Climate Conundrum: Why Didn't Atmospheric CO2 Fall During the Pandemic?" *Caltech Magazine*, Spring. Accessed November 23, 2022. https://magazine.caltech.edu/post/atmospheric-co2-covid-pandemic#:~:text=Notably%2C%20emissions%20returned%20to%20near,cutting%20emissions%2C%20the%20study%20noted.

Sisco, Matthew, Silvia Piana, Elke U. Weber, and Valentina Bosetti. 2021. "Global Climate Marches Sharply Raise Attention to Climate Change: Analysis of Climate Search Behavior in 46 Countries." *Journal of Environmental Psychology* 75. doi:10.1016/j.jenvp.2021.101596.

Thorbecke, Catherine. 2022. "Experts Slam Oil Giant ExxonMobil's Net-Zero 'Ambition.'" *ABC News*, January 19. Accessed November 22, 2022. https://abcnews.go.com/Business/experts-slam-oil-giant-exxon-mobils-net-ambition/story?id=82325190.

WMO. 2021. "Weather-Related Disasters Increase Over Past 50 Years, Causing More Damage But Fewer Deaths." *WMO*, August 31. Accessed November 24, 2022. https://public.wmo.int/en/media/press-release/weather-related-disasters-increase-over-past-50-years-causing-more-damage-fewer#:~:text=According%20to%20the%20WMO%20Atlas, US%24%203.64%20trillion%20in%20losses.

Yergin, Daniel. 2020. *The New Map: Energy, Climate, and the Clash of Nations.* New York: Penguin.

Index